9.95

SEP 87   29 '95

REG.
PRICE

STRAND PRICE
$1.00

TECHNOLOGY, ECONOMIC GROWTH
AND THE LABOUR PROCESS

TECHNOLOGY, ECONOMIC GROWTH AND THE LABOUR PROCESS

Phil Blackburn,
Rod Coombs
and
Kenneth Green

St. Martin's Press   New York

ISBN 0-312-79001-5

Library of Congress Cataloging in Publication Data

Blackburn, Phil.
  Technology, economic growth, and the labour process.

  Bibliography: p.
  Includes index.
  1. Technological innovations. 2. Technological
innovations—Social aspects. 3. Economic development.
I. Coombs, Rod. II. Green, Kenneth. III. Title.
HC79.T4B55 1985   338'.06   84-22849
ISBN 0-312-79001-5

To Barbara Wilkinson, 1935–84

Contents

Contents

Tables

Figures

Preface

Being the product of three writers, and ranging as it does over wide expanses of literature in the fields of industrial sociology, political economy and the history of technology, this book has relied on many people in its preparation. The three authors have worked together, in various combinations, on some of its themes from the late 1970s, first in the Department of Science and Technology Policy at the University of Manchester where one of us (Kenneth Green) still lectures. Subsequently, we have continued our work on the 'political economy of technological change', one of us (Rod Coombs) as a lecturer in innovation at the University of Manchester Institute of Science and Technology and the other (Phil Blackburn) as a researcher in industrial policy at the Economic Policy Group of the Greater London Council. In all these institutions, we have discussed the issues we raise here in classrooms, seminars and meetings and we would like to thank all those who have helped us to formulate our ideas. In particular we would like to thank those with whom we have worked and on whose writings we have drawn in various chapters of the book – Julian Cohen, Jeff Atkinson, Alan Bornat, Heather Aspinwall, Nick Higgett and Penelope Whitehead. Especial thanks to Elena Softley who has allowed us to use extracts from her thesis on word-processing and office organisation in our discussion of neo-Fordism in clerical work in Chapter 7.

Many of the ideas have been aired at conferences, in Britain and abroad, particularly at various international seminars on long waves, annual meetings of the Conference of Socialist Economists and, in 1981–2, meetings of the Group for Alternative Science and Technology Strategies. We would like to thank Chris Freeman, Bryn Jones and Sonia Liff who have, in one or other of these forums, helped us to a better understanding of the political, social and economic dimensions of technological change. Viv Walsh, Jon Turney and editorial consultant Keith Povey are to be thanked for their useful comments on the final draft, and the authors are indebted to Yvonne Aspinall, Collette White, Gill Greensides, Maggie Soltani and Barbara Wilkinson (who tragically died during the final stages of the book's preparation) for typing various parts of the seemingly never-ending stream of drafts and redrafts. Of course, only the authors are responsible for any errors and confusions in what follows.

PHIL BLACKBURN
ROD COOMBS
KENNETH GREEN

Manchester

Acknowledgements

Acknowledgement for permission to reproduce material from the following books and articles is gratefully made to: Conference of Socialist Economists (for Table 1.1); New Left Books (for Table 4.1); Christopher Freeman, John Clark and Luc Soete (for Table 4.2); National Productivity Review (for Tables 6.1 and 6.2); Jonathan Gershuny and Ian Miles (for Table 7.1); Harvard University Graduate School of Business Administration (for Figure 2.1); Engineering Industry Training Board (for Figures 2.2, 3.1 and 3.2); William Heinemann Ltd (for Figures 6.1 and 6.2); Financial Times Business Information Limited (for Figure 6.4). Every effort has been made to trace the original source of material in this book. Where the attempt has been unsuccessful, the publishers will be pleased to hear from the copyright holder.

1 Introduction

INTRODUCTION

Public interest in technology appears to move in waves. From World War Two to the late 1960s technological developments – particularly prestigious ones like atomic power, space technologies and 'automation' – were embraced by people of all political persuasions as essential contributors to a rapidly growing economy. By the late 1960s this enthusiasm had become tempered by a concern for the undesirable aspects of technology. This was stimulated by the appalling demonstration of technologically-based American military might in Vietnam, by the massive spending by rich states on prestige projects (particularly in aerospace) in a world still riven with hunger and poverty and by the accumulation of a host of environmental problems in air, water and land produced as side-effects of post-war economic growth.

The 1970s were a time of ecological and political critiques of scientific, technological and medical developments. Expensive programmes in aerospace and nuclear technologies were run down, after substantial political struggles, and new agencies were established with more stringent guidelines to regulate industrial processes and consumer products. But from the late 1970s, coinciding with the arrival of a deep depression in economic activity, interest in technology has risen again – focused this time not on the costs and benefits of large-scale technologies but on developments in microelectronics and its applications, computer-assisted production processes, robotics, telecommunications and information technology.[1]

In all advanced capitalist countries these developments have become the subject of substantial political argument as governments seek to harness them to their programmes for economic recovery. Massive investments by private and public telecommunications organisations are in train to provide, so they hope, the infrastructure for an information technology 'revolution' which will stimulate new markets and offer new profitable areas for firms emerging from the

1

slump. Indeed micro-electronics and information technologies, in their entirety, are seen as 'historic' developments, whose significance in their effects on economic and social structures will turn out to be as important as were the application of steam power in the Industrial Revolution, electricity in the late nineteenth century and auto-mobiles, aeroplanes and plastics since the 1920s.[2]

Much of this is exaggeration; there are not enough years left in this century in which to cram all the predicted technological revolutions. Nevertheless it is plain that these new technological developments, fixed into new manufacturing processes, into new products and into new modes of delivery of producer and consumer services are involved in some important changes which are taking place. These are changes that are seen in the industrial structure of all countries, in the organisation of production and of work, in the amount of work to be done, in patterns of individual and collective consumption of products and services and, in short, in the way people communicate with each other and live their lives.

This book addresses itself to one aspect of these technological developments, namely, what is the significance of current technolo-gical developments in the context of the prevailing economic crisis and how can 'new technologies' fit into any economic recovery? This question becomes a family of related questions. What changes in the organisation of production of physical products and of services 'go with' new technological developments in such a way that the higher levels of productivity may be realised? Or, what changes in the industrial structure of advanced capitalist countries would seem to be strategically necessary and can be detected in their initial stages, if the opportunities for profitable investment in new technology-based products and services are to be realised by capitalist firms? Or again, what institutional innovations, associated with the use of new technologies as an element in the restructuring of production, are open to firms and to governments managing mixed economies in any incipient period of economic recovery?

By discussion and elaboration of which bodies of ideas are we to examine these issues? Most of the literature on the so-called 'effects' of new technologies, a literature which has exploded over the last five years, is of little help. It is frequently naive in its estimation of the speed with which the new technologies will move from the drawing-board to real-life economic applications or of the rate of diffusion of those new technologies throughout a particular industrial branch. In short, if it does not exaggerate the speed of change it is insensitive to the national and international social and economic problems and adjustments which any supposed rapid technological change would bring (Green, 1984).

So rather than looking for insights into likely techno-economic trajectories of the next twenty years from the 'impact of new technology' literature, we draw on two bodies of literature which are somewhat more substantially founded and which we bring together, we hope, in a new and fruitful synthesis. The two are: labour process theory and long-term economic development theory. We try to place new technologies currently being developed or innovated into a broader context of historical development of capitalist economies, synthesising the accounts of long-term economic development and the analysis of technology developed by historians and writers on the capitalist labour process.

LABOUR PROCESS ANALYSES

'Labour process' writers have offered Marxist accounts of the workings of the capitalist mode of production which focus on those combinations of human labour and technology in particular organisational settings which comprise the heart of production. Over the past ten years, the touchstone for these analyses has been the book *Labor and Monopoly Capital: the Degradation of Work in the Twentieth Century* by Harry Braverman, published in 1974. We do not intend to give any extensive account of the contributions to an understanding of production process developments which labour process writers have made, as there are already several reviews available (Wood, 1982; Thompson, 1983). We will be using the work of many of these writers in this book. Nevertheless some general remarks are necessary.

Labour process theorists have focussed attention on the relationship between two sets of motivations which induce managers to change production technologies. One set of motivations concerns the conventional calculus of production costs. The other set of motivations concerns the relative power of workers and managers to exercise control over the manner and pace of work. The labour process literature has tended to polarise into two interpretations of this problem. One stance maintains that from the time of the transition from 'formal' to 'real subordination' in the labour process, capitalist economies have displayed an intrinsic tendency to remove the control of all the labour process from workers.[3] It is argued that this appropriation of control by managers creates a 'deskilling' bias to the evolution of production technology. This position does not necessarily ignore the cost motivations; it simply emphasises the control motivations. The contrasting position emphasises the negotiable character of the jobs and skills which accompany any particular machines introduced into an industry. The same machine tool, for example, will generate different skills shifts when introduced into different factories, dependent upon the other specific factors present in those workplaces.

These labour process analyses, though extremely fruitful in their focus on the relations between workers and the technological system, and between different sections of workers themselves, are of limited usefulness if they concern themselves only with the relationship of control and resistance between management and workers on the shop-floor at the expense of examining the wider context of changes at the level of the political economy. Those arguing the negotiable character of any labour process change and who are against seeing deskilling as a dominant tendency in the development of the capitalist labour process, still remain within that logic. Tendency and counter-tendency, one managerial strategy or another, old or new technology, the focus is the same – the control of labour.

Much of the literature has thus lacked an appreciation of the changes in labour control systems in the broader context of changes in production systems as a whole. In particular, there has been a tendency to see technology as either given, or else entirely shaped by the needs of a particular division of labour (which is either determined by management or by interaction between management and workers). Without entertaining technological determinism, it is possible to identify some constraints and incentives in the development of production systems and the *technologies* within them which are not entirely reducible to the explanatory variables employed by writers on the labour process. In other words, it is possible to identify sequential shifts in the development of labour processes which occur with *discontinuity,* rather than with necessary continuities as assumed in labour process writings. This task is attempted in this book and the results are found to be related in an interesting way to the phenomenon of long waves in the development of the world capitalist economy.

We should stress that we are using 'labour process' as a term referring to the combinations of human labour (in all its forms of manual and mental skill in the conception and execution of productive tasks) *and* machinery/tools/infrastructures ('technology') which are necessary to produce some physical or service product from a set of raw materials (which may be the products of other equally complex labour processes). We stress this *combination* of the two elements of labour and technology because we observe a tendency in many labour-process analysts, eager to put human activity in its obviously vital place in production, to marginalise the technological element. Whilst such emphasis are important as a correction of the bias of some industrial sociologists and management consultants who place undue emphasis on the so-called technological imperative, they must not be overdone. One of us (Coombs, 1984) has suggested a return to a general concept of a 'production process' instead of 'labour process'. This might avoid an over-emphasis on trends in the use of human labour in explanations of developments in production. In this book however, we have continued to use the labour-process concept because of its important ideological function in reminding us that labour is a vital element of production. We use the terms 'labour process' and 'production process' interchangeably throughout this book.

LONG WAVE THEORIES

As the economic crisis in the advanced capitalist countries has deepened, increasing interest has been shown in a set of theories of long-term economic development, called long-wave theories, which emphasise changes in the structure and organisation of industrial sectors which accompany successive phases of economic growth. Some writers have tried to associate the waves with leading industrial sectors consisting of products and processes which have played a dominant role in the economic expansion of the developed capitalist countries. Table 1.1 gives one version of this, showing the dates of the waves' upswings and downswings. The long-wave literature is now quite large and the full range of theories cannot be fully reported in this book. Of those theories exploring the significance of technological change as a factor in the upswings and downswings of the long wave, those of Freeman, Clark and Soete (referred to in this book, in shorthand, as Freeman) and of Mandel are the most substantial. (For a review of the principal theories, see Delbeke, 1981.)

TABLE 1.1 *Long waves and associated industries (illustrated for Britain)*

Dates	Industry	Technological Developments
1790s–1820 (upswing)	Cotton	Mechanisation of spinning
1820–mid-1840s (downswing)		
mid-1840s–1870s (upswing)	Textiles	Mechanisation of spinning and weaving
	Engineering	Production by machine of textile machinery, steam engines and locomotives
1870s–late 1890s (downswing)		
late 1890s–1920 (upswing)	Engineering	Batch production of semi-automatic machinery; marine engineering; motor car
	Electrical Chemical	} Rise of science-based industries
	Steel	Bulk production
1921–World War Two (downswing)		
World War Two–early 1970s (upswing)	Motor cars, Mechanical and electrical consumer durables	} Assembly-line mass production
	Petrochemicals	Continuous flow processes
early 1970s–?		

SOURCE   *adapted from Conference of Socialist Economists Micro-electronics Group (1980).*

In the work of Freeman *et al.* (1982) the starting point for the description of the long wave is a rejection of the hypothesis (put forward by Mensch, 1979) that a cluster of technological innovations is created in the depths of an economic depression. Following the pre-war political economist Schumpeter, who wrote penetratingly on the last major capitalist economic crisis, they argue that many innovations are in fact serendipitous. What matters for *long waves* are surges of capital investment and the *diffusion* of technological innovations. Freeman proposes that the upswing of the long wave is characterised by the roughly simultaneous diffusion of a number of technically-related new products and processes with their associated new firms or branches of firms; they call this a 'new technology

system'. In Freeman's view, these new sectors of the economy have
the ability to tip the balance of the macro-economic system from
prolonged stagnation and crisis into an expansionary, full-
employment growth path, with a virtuous circle between productivity
growth and demand growth. Secondary booms in consumer durables
fuel the process once it is under way. This role is attributed to those
industrial sectors which exploited the technological developments in
electronics and synthetic materials for the post-war long-wave
upswing. Eventually however, this impetus is seen as necessarily
subsiding.

A number of factors are responsible for the slowdown. Market
growth in new sectors eventually encounters limits, monopoly profits
are competed away, capital and labour market rigidities create
inflation in the long run and institutional responses result in some
deflation. In the absence of a second 'new technology system' the
combination of declining growth rates and increasing inflation leads
to stagflation and eventually to recession. Throughout this process
Freeman suggests that the increasing cost pressures and the increasing scale of the new sectors will gradually create a labour-saving bias
to technical change. He therefore sees mechanisation as a phenomenon which is intensified by the recession, but he does not specify
what technological form it might take, nor whether there is a
connection between mechanisation, capital goods innovation and the
new technology system implied in the long-wave mechanism.

Freeman's account offers an explanation of the progress of a long
wave from early upswing through an upper turning point, and into
downswing, but it does not directly confront the issue of the lower
turning-point except to invoke 'socio-political factors' without further
elaboration. In common with several writers (Perez, 1983; Dosi,
1983) he sees the extremes of social distress and conflict which occur
in the downswing as capable of triggering institutional changes which
permit radical change in the structure of economies and a return to
long-term economic growth.

Mandel's analysis (1975, 1978, 1980) is somewhat different to that
of Freeman's . It is firmly in the tradition of Marxist economics and a
substantial part of the proposed mechanism is a reinvestment cycle
analogous to that proposed by Marx for the short trade-cycle. Mandel
however suggests that major technological revolutions in the capital
goods industries periodically occur, and that they cannot be
accommodated in the period of a short cycle. Therefore over a
number of short cycles a reserve fund of 'under-invested' capital

builds up which eventually forms one of the stimuli to a long-wave upswing. The major stimulus to restructuring however is the prospect of long-term profit rates being favourably influenced by changes in socio-political conditions. Therefore Mandel is also making the lower turning-point of the wave historically contingent, but within a class-political framework of analysis. The technological revolutions which, in his view, coincide with the three most recent long-wave upswings (1850–76, 1896–1914, 1945–1967) are steam-powered production of steam-powered machines, electrification and the internal combustion engine and 'automation'. By this last term, Mandel means mechanical handling, continuous flow production and electronic and computer control.

'AUTOMATION' AND FORDISM

So, a number of long-wave theorists explain the boom in capitalist countries in the 1950s and 1960s principally in terms of the rise of new industries based on 'new technology systems' and/or on the spread of what they call 'automation' (which, as we will explain, is a singularly unhelpful term). In our view these explanations are only partially satisfactory, for two reasons.

First, they display an inadequate understanding of what 'automation' up to now has actually involved, seeing it as some qualitatively new form of production organisation clearly distinct from previous forms extant up to the 1930s. One essential element of automation, namely the mechanised transfer of items under manufacture between the machines or assembly workers (as in the conveyor line) had been in use in some industrial processes since the end of the nineteenth century. Another essential element of automation, the mechanisation or computerisation of the human control aspects of any production activity was, to say the least, in its very early stages in the 1950s. Indeed, as many information technology pundits declaim, it is only *now* with the application of cheap microcomputers to industrial and service production processes alike that the majority of production sectors, including those like small-batch engineering and office work relatively untouched by the massive productivity advances of the last boom period, are susceptible to 'automation' in the fullest sense of the term.

Second, insufficient attention is paid to changes in non-technological aspects of the production process which significantly facilitated the last boom. The successful introduction and diffusion of any new technology system requires some related radical reorganisation of the non-technological parts of the production process. As labour-process theorists have pointed out, the application of the particular technological systems that were implicated in the last boom required a particular form of labour discipline, division and management which had been created in the high unemployment and slump conditions of the inter-war years. Without the prior establishment of a labour force accustomed and willing (because it had no option) to take up the unskilled jobs of assembly and packing in mass production industries or suitably technically-trained in the complex tasks of product and process design, production planning and management, then no mass-production based, employment-generating boom was possible.

The past fifty years have seen phenomenal changes in the structure of production and consumption, both in terms of the nature and quantity of the physical goods and services produced and consumed and in terms of the productivity and the manner of reproduction of the labour they require. These changes are the result of a variety of technical and economic developments. New industries have arisen by exploiting the products of investment in scientific research. Highly mechanised systems of production have come to many industrial sectors; now new capital equipment is clearly poised to enter many others, even those sectors (like banking or retailing) not traditionally associated with machinery. Less often emphasised, but in our view as important as the other two, has been the rise of a 'paradigm' of production organisation in mass-production industries based on product standardisation and the large-scale use of unskilled labour for final assembly. This was a model of how production should and could be organised and acted as a touchstone of productivity comparison even for those industries and services which are unable to adopt it fully, either because of the nature of their commodities or the structure of their markets.

Furthermore, without markets and the *consumption* patterns of the population being geared to the large-scale purchase of mass-produced goods, an historically new situation, then mass *production* would make no economic sense. In short, any explanation of the last boom period needs to include some understanding of the development and diffusion, industry- and world-wide, of certain patterns of production organisation and of consumption of goods and services which we call *Fordism*.

STRUCTURE OF THE BOOK

Chapters 2, 3 and 4 explore the developments in mechanisation and the concept of Fordism further. Chapter 2 re-examines and redefines the concept of automation; Chapter 3, drawing on the writings of various historians of technology and of labour-process theorists, reviews the history of production organisation over the last hundred years. Chapter 4 explores the role attributed to technological change by various long-wave writers and presents some empirical evidence on the diffusion of automation in the USA and in Britain since World War Two. Connections are observed between the various phases of mechanisation and of production organisation and the upswings and downswings in long-term economic development. In these three chapters it is argued that the historically observed technological change in production is best seen as a series of stages which have in common an increase in the complexity of mechanisation. We identify three phases of mechanisation, which we call primary, secondary and tertiary mechanisation. Production technology however is not simply hardware. It is intertwined with forms of human control, co-ordination and planning in specific labour processes which, in capitalist production systems, are governed by the need for a generation of profit. Fordism, it will be argued, is that particular combination of technology and human labour which has broadly corresponded with the elaboration of the secondary mechanisation stage of technological development.

It is reasonable to ask whether the current economic crisis, the recession and depression phases of the long wave as some would say, is associated with some 'exhaustion' of the Fordist paradigm which has now reached some limit. Is Fordism appropriate for the successful application of information technologies within those sectors of production which have hitherto not benefited from it – for example, small-batch engineering and office work? These are the issues we will address in this book. We argue that the post-war boom can be characterised as containing the combination of a specific technological form (namely maturing secondary or transfer mechanisation with emergent tertiary or control mechanisation) and a specific system of labour organisation – together comprising Fordist production processes. Any new boom will contain a new combination of those elements.

Technologically, it is reasonably clear what will be one of the bases of the new combination, namely the use of micro-electronic-based and computer-dependent information technologies, though exactly what these constitute needs to be delineated. The likely forms of labour process are more difficult to ascertain. Nevertheless there are enough 'examples' of labour organisation around tertiary mechanisation to allow some informed speculation. In Chapters 5, 6 and 7 we explore two industrial sectors in which information technologies will loom particularly large over the next twenty years, as new computerised control systems permit large increases in labour productivity by 'mechanising' highly labour-intensive activities of information flow and the co-ordination of production. Chapter 5 reviews recent literature on work redesign and seeks to identify an emergent form of technological/labour combination we call (in the style of Palloix and Aglietta) *neo-Fordism*. Chapter 6 looks at technological and labour organisation methods in small batch engineering. Various writers have looked at the changes in labour organisation following the introduction into small-batch engineering of computerised NC-machine tools. They have seen the changes in that industry as only partially successful (if not very *un*successful) attempts to shift small-batch engineering from craft to Fordist systems of production organisation. We see them as partially successful attempts at a shift straight to neo-Fordism, a shift now facilitated by the availability of such tertiary mechanisation as flexible manufacturing systems and computer-aided management.

Chapter 7 takes a wider focus, examining Fordism's limits not just as a form of production organisation but as a form of the structuring of consumption, limits which would need to be crossed if capitalist economies were to revive themselves by an injection of information technologies as components of production processes and mass consumer products. We first explore the application of information technologies in clerical services and, as in Chapter 6, examine how the labour processes of clerical work are likely to be transformed, elaborating on the concept of neo-Fordism. However, the longer-term potential of information technologies is in a complete *restructuring* of many industries in the service sector which traditionally have been of low capital intensity and certainly have had little mechanisation. As we will discuss, some writers are of the view that some service industries – particularly in the welfare sector of health care, education and public administration – are in urgent need of some reorganisation if their high costs are to cease being a drain on capital accumulation. But any reorganisation of these services requires a change not just in the labour processes of providing them but also in the services themselves, that is in the manner in which they satisfy the human needs for which they exist. We explore this consumption-restructuring aspect of neo-Fordism by looking at what is going on, and what information technologies offer in the longer term, in health care provision.

We should say that although this book focuses on current technological developments, we are not trying to suggest that those developments will *inevitably* force changes in production organisation that will be the basis of a resurgence in economic growth in capitalist economies. The questions we raise about the role of technological developments in the process of restructuring of depressed capitalist industries seem important to us because they indicate the *strategic options* in the use of new technologies which might be presented to capitalist firms and governments. We agree with those long-wave writers mentioned earlier, that institutional and political changes have to accompany technical and economic change in order to return to economic growth. But a closer analysis of the historical development of production systems, of the type we attempt in this book, is an important gap which needs to be filled. It is needed to reveal and clarify the ground upon which political struggles over new technology will be conducted. Chapter 8 reviews the general arguments of the book and points to some of those issues, suggesting where alternative criteria to those of the market alone might find some space.

2 On New Technologies and 'Automation'

INTRODUCTION

The last few years have seen the introduction of several new technologies into industry and commerce which are said to have 'automated' many previously labour-intensive production activities. Yet such automation has not in any simple way been the cause of the high level of unemployment which the advanced capitalist countries are experiencing in the 1980s. That unemployment, although partly exacerbated in some countries by demand deficiency as a direct result of government policies, has structural causes. The slump of the early 1980s needs to be explained in terms of these structural features of the world economy before labour-displacing technological change can be satisfactorily analysed.

Slumps are a recurring feature of capitalist economies, and they have a 'function' which bears upon technological changes. During slumps, competition intensifies for shares of declining, static or slowly growing markets in many products and services. This may result in the weeding out of weaker firms, forcing rationalisation and industrial concentration which nowadays is likely to be on a world scale. This process necessarily implies the shedding of labour by some firms and some workplaces in many countries. The rationalisation may involve more efficient use of machinery and systems of production which make workers work harder and/or longer (what Massey and Meegan, 1982, call 'intensification'). Often, however, it involves a change in the products being made to satisfy changed consumer demand or to increase the quality of the product or service. Furthermore, there may be redistribution of productive activity within a company, either within one country or within transnational corporations (what Massey and Meegan call 'rationalisation'). These activities involve technological change in a variety of forms. During the rest of this decade there will be further refitting of production

13

processes – in both manufacturing and services – with new equipment. Likewise there will be further opening up, often in completely new geographical locations, of new factories, offices, etc., based on radically new technologies.

But, although new production technologies have not as yet resulted in much job loss, because all the other rationalising agents have been at work on a much bigger scale, this is not going to continue. Radically labour-displacing technologies are still in the introductory phase of their diffusion curves. They are about to take off. As firms invest in the 1980s – and slump or no slump, investment in new factories and new equipment still takes place – the machines and processes they install will not require as many workers for the same level of output as the older equipment that they and their competitors are scrapping. Much depends, therefore, on the opportunities for expansionary, rather than rationalising, investment, and on the other factors which condition those opportunities.

Although classification is difficult, there are five main types of civil technological development which together might be held to constitute that 'new technology' which is expected to have considerable economic and social significance over the next two decades, although the degree of 'newness' of each type may be disputed. The five types are: micro-electronic-based technologies, information technologies, biotechnologies, new materials and energy technologies. (Elsewhere we have called these 'Chips, Satellites, Bugs, Nodules and Windmills'.)[1]

New energy technologies and new materials could clearly be of importance for any resurgence of economic growth, given the substantial changes in raw material and energy costs over the past fifteen years and the increasing emphasis on conservation of resources and reduction of pollution.[2] By biotechnologies we mean the genetic engineering of all sorts of biological species to produce new strains of plants and animals for agricultural use and of bacteria and fungi for use in the manufacture of drugs, chemicals and foodstuffs. Such developments may transform the chemical, pharmaceutical, energy, agricultural and food-processing industries, although not very quickly. Biotechnology's economic importance is likely only to become large at the beginning of the next century, mirroring the importance of the chemical industry in this one.[3]

But the economic impact of micro-electronics-based and information technologies is already very apparent and has been the subject of considerable study over the past five years. 'Microchips' have already been incorporated in numerous industrial and consumer products and this will continue at a rapid pace. The reduction in the number of metal-worked electro-mechanical components in domestic consumer products (washing machines, typewriters etc.) and their partial replacement by low labour-content micro-electronic components, reduces the need for metalworkers to make the now obsolescent metal parts and also reduces the number of assembly workers required. When a new product is being designed, it will become routine, as it has become in the watch and calculator industries, to assume that the product should incorporate the smallest number of standardised electronic components requiring much less labour in assembly.[4]

Particularly important for any analysis of trends in automation is the development of cheap and versatile computer systems which can be incorporated into a variety of industrial and service production processes. Together with other components of so-called 'information technology', namely office machinery and telecommunication systems, these will make up the largest part of new technology investment of the next ten years. It is on these types of new technology that this book focuses.

NEW TECHNOLOGIES

There are four sorts of equipment which will be the basis of this new technology investment: robots, manufacturing process computerisation, office machinery and the 'electronicisation' of information-handling.[5] We will take each of these in turn.

Robots

The application of robots to routine handling, assembly and packing activities will speed up over the next ten years. As yet, contrary to the association in the popular mind of automation with robots, very few robots have been introduced into industrial production, displacing very few workers as a result. At the end of 1983, there were about 40 000 robots in productive use in the advanced capitalist countries (including 16 500 in Japan and 1750 in Britain).[6] Most of them were used for simple jobs like spot-welding (in the car industry, for example) with painting, injection moulding handling, arc welding and transfer between machine tools making up the rest. Their biggest *potential* use is obviously in labour-intensive mass production assembly and inspection of light industrial and consumer goods and in packing awkward shapes (for example, putting chocolates in boxes). It is on these applications that robot engineering development is now concentrating.[7] We might expect that by the early 1990s, insofar as products are re-designed so that they can be more readily assembled by cheap, industrially-proven robots, the world robot population will have increased tenfold (OECD, 1984). This must mean that, other things being equal, the number of jobs in unskilled assembly and packing will decline at an increasing pace over the next ten years. This will have a particularly serious effect on women's employment in these jobs (SPRU Women and Technology Studies, 1982).

Manufacturing Process Computerisation

The growing sophistication of computers means that they can be increasingly incorporated into processes and into individual machines that so far have been heavily dependent on human labour for control. Even when an industry is already highly mechanised (for example, chemicals) the scope for computer applications in the production machinery may still be high. Computerisation of the manufacturing process control system may increase machine speeds and throughput and the quality of the product can be improved by more rapid and accurate inspection and correction of deviations in the process. Since micro-electronic control equipment is also likely to be very reliable, significant increases in the productiveness of even already highly 'automated' machinery can be expected. However, while the technological scope in this type of capital-intensive process industry may be high, the change in the number of people required to operate the machinery is likely to be small, since the existing high level of mechanisation means that the labour input overall is already relatively low anyway. So, in industries like chemicals, oil-refining and some parts of textile yarn manufacture where continuous processes have been introduced systematically over the past thirty years, the inclusion of computers in new processing technology might have significant effects on productivity but without radically affecting the numbers employed or the skills expected, at least as far as the workers directly involved with the continuous process itself are concerned.[8]

There are, however, a number of industries in which the application of cheap computer control systems offers the potential for turning them into capital-intensive process industries. This has been underway in the steel, food and printing industries for some time.[9] It is beginning in coal-mining in which the mechanised cutting and conveying of coal to the surface are being integrated under computer control.[10] It is also happening in textile finishing. Although very considerable amounts of machinery are already used in dyeing and finishing textiles there has always been the need for skilled human intervention to assure the quality of batches of finished cloth. Computer-based textile-finishing control equipment makes up a new generation of textile process control which will have considerable effects on the number of human quality controllers in that industry (see Green *et al.*, 1980).

Even where the nature of the product being manufactured makes fully integrated computer-controlled process systems inapplicable, there is still scope for the introduction of individual computer-controlled machines. An example might be the programmable sewing machine in clothing manufacture (McLean and Rush, 1978; Braun and Senker, 1982). However, of more significance for economic change in the longer term is the introduction of such machines into the small-batch engineering industry. Small-batch production, as distinct from the continuous or large batch production of one good, has been an obstacle to the introduction of integrated processes. In general, heavy investment in 'automated' equipment cannot rapidly be recovered in small-batch production given the time taken up between runs in re-setting machines or even re-organising the factory layout. Also, small-batch production introduces constraints into the operation and development of production machinery.

It is a widely-held idea that computer-based machinery such as computer numerical-controlled machine tools could rapidly be developed so as to produce routinely very small batch pieces to any specification – using 'Flexible Manufacturing Systems'. It has long been the dream of designers of small-batch engineering manufacturing systems to be able to use computers to monitor production as a whole, in real time. The result would indeed be to turn complex, non-routine small-batch engineering production into a sophisticated process-type industry. Computer-aided design and draughting could be linked with computer-controlled machine tools in systems of computer-aided manufacture. Reductions in the size and cost of computers and an increase in their power will add impetus to the development and diffusion of these techniques but this process will still not be fast, because of the significant technical and financial constraints. The impact on labour required as a result of the use of such systems in small-batch engineering would, of course, be very high. Such systems however, although under development in various countries, are still some way from being economically viable. Nevertheless, as we discuss in Chapter 6, they are important in understanding the strategic options open for economic recovery in the next twenty years.

Office Machinery

The diffusion of electronic office machinery is advancing rapidly. In the late 1970s and early 1980s, world sales of electronic word-processors were growing at a rate of 25 per cent a year (Werneke, 1983). Other equipment, like photocopiers and microcomputers, were also diffusing into offices at a rapid rate. The use of such equipment, it has often been predicted, would lead to very large reductions in the number of secretaries, typists and other clerical workers (see Bird, 1980). Although word-processor-induced job losses have been partly concealed by the general recession, a recent survey showed that about a quarter of British offices using electronic office equipment have reduced the number of typists they employ; less than 3 per cent increased the number (Steffens, 1983). Word-processors are, of course, only the first stage in the computer-based mechanisation of office work. The 1980s are forecast to see the growth in the number of integrated electronic offices. As the survey reported, a third of the offices surveyed (mainly the smaller organisations) were buying their electronic products as single, stand-alone items, and a third (mainly the larger ones) saw them as the initial part of a full electronic office system. The rest were waiting to see how things went with their first purchases before deciding whether or not to go ahead with a fully integrated system.

The principal reasons for investing in electronic equipment were that it enabled quicker text-processing, better quality of presentation, better services to customers and to management and, of course, savings in labour. So even though much office work is not amenable to substantial computer-based 'mechanisation' there seems little doubt that the introduction of electronic office products – from word-processors to facsimile machines, and executive work stations to electronic mail – will speed up over the next few years and as a result have major effects on the structure and size of office employment.

'Electronicisation' of Information-handling

The development of new electronic-based telecommunication and data transmission systems means that a wide range of information-handling service industries are likely to be transformed, as labour-intensive clerical activities are computerised. Banking might serve as an example here. The volume of business in banking has grown tremendously in recent years and could continue to grow as an increasing proportion of the population acquires bank accounts (particularly in Britain.) Employment in banking has also grown considerably, but less quickly than the volume of business. The already considerable use of computers is the principal reason for this productivity growth. The basic trend seems likely to continue, with several possible areas in which computers can contribute to a slower increase in employment than in output. Automatic tellers (for example, cash points) will become more common and reduce the labour requirement in counter operations. It is forecast that there will be at least 70 000 automatic tellers in the USA by 1990, compared with the 20 000 in 1980 (Werneke, 1983). The banks are in the vanguard of developing Electronic Funds Transfer Systems, replacing paper money for many transactions; automatic cheque-clearing systems may be introduced; there may be a tendency to create 'two-tier' organisational structures in banking with High Street branches dealing only with counter operations and regional centres handling administrative jobs and specialist banking tasks, such as industrial loans.[11] Such reorganisation would facilitate the use of electronic office equipment in the upper tier of the system. As Sleigh *et al.* (1979, p. 67) conclude:

> The net result of the various trends in both business and technology which are currently facing the banks seems likely to be that modest increases in employment are likely to continue for at least the next five years and that stability in numbers will be achieved sometime between 1985 and 1990.

Similar developments, linking computers and other office machinery with new communications systems so that the handling of information becomes less paper (and people)-based and more electronic-machine-based can be envisaged for other service industries.[12] The speed of diffusion of such equipment is very much dependent on the availability of the inter-computer communications equipment – the telephone lines, data transmission cables, satellites, microwave transmitters and so on. Such extensive infrastructures are currently planned or being installed in every advanced capitalist country, so it is clear that in the longer term, information technology has tremendous potential for reorganising the way services are provided to other industries and to direct consumers. (These are discussed in more detail in Chapter 7.)

Conclusion

So far, in discussing the sorts of new technology investment likely to have economic significance over the next ten years, we have emphasised the job-destroying potential of such 'automation'. Of course this is not the whole story. At the same time as existing industries are being rationalised and the new productivity-increasing investments are being made in new processes, new factories and offices are opening to produce completely new products which are in effect creating new jobs (though not necessarily with the same skills as those being lost or in the same part of a country or of the world). As in any slump, job-destruction is accompanied by job-creation and what determines long-term employment prospects is the balance between the two in this process of *restructuring*. However, this book is not specifically concerned with the debates over employment issues concerning new technology, rather with the contribution of new technological developments to any upturn in the developed capitalist economies. It is assumed, particularly by writers on long waves in economic development, as we mention in Chapter 1, that any upturn will involve the application of new technologies to increase the productivity of labour – to increase 'automation'. Having outlined the kinds of technological development that might be involved in 'automation' in the 1980s, we turn to a more basic analytical question: what exactly *is* the phenomenon which 'automation' is supposed to describe? As a term, does it help us to understand or does it obscure the processes of technological change?

WHAT IS THIS THING CALLED 'AUTOMATION'?

Introduction

The term 'automation' was coined in the 1940s, and the breadth of definitions since assigned to the word, coupled with its widespread entry into everyday language, has made it of limited use as an exact description of any well-defined phenomenon. On the contrary, it has become common parlance for all the frequently observed ways in which machinery replaces or appears to replace human activity. It is for this reason that we have so far put the term in inverted commas; we will continue to do so throughout this book.

Ever since people have become concerned with the efficiency of generating products or services they have constructed their notions of efficiency in terms of the saving of labour input necessary for a given output. At various times in history, concern over the favourable or unfavourable consequences of such increases in efficiency has become more pronounced. In recent years there have been two such periods. The first of these stretched from the early 1950s to the early 1960s and was concentrated primarily in the USA, though with significant echoes in Europe. The second began in the early 1970s.

Early Discussions

The literature of the first period has been surveyed by Cohen (1975). The significant features of that period of discussion can be summarised under three heads, concerning *definitions, job loss* and *economic growth*.

Definitions

There was no agreement on how 'automation' could be defined. Cohen identified over thirty definitions from the writings he studied. They vary from the broad plain language type to the more precise but strongly contested. They were notable for the fact that they generally fell somewhere between the language of engineering and the language of management or economics, as a result of trying to capture the notions of machines substituting for people and increased labour efficiency. At the most general level, some authors refused to differentiate 'automation' from technical change in general; for example: 'Broadly, automation now encompasses *practically any device* that reduces the amount of human effort – physical, mental or both – necessary to do work' (O'Brien, 1969; emphasis added). Slightly more specific were those who labelled it as the 'latest phase of mechanisation'; for example: '[Automation is] any continuous and integrated operation of a rationalised production system which uses electronic or other equipment to regulate and co-ordinate the quality and quantity of production' (Buckingham, 1955). A more specific type of definition was used by those who spoke of 'Detroit automation', by which was meant the mechanised transfer of workpieces (engine blocks) from one work station to the next, characteristic of the USA automobile industry from the 1940s. One generalised version of this last definition is contained in the definition of 'automation' as 'systems engineering', which implies large scale co-ordination of workflow as a prominent design imperative in the construction of factories. Finally, but perhaps more frequently, the ideas of feedback and control, particularly achieved by means of then emerging electronic technologies, constituted another style of defining automation.

Clearly, there are many overlaps between these definitions, and in many cases the same author uses different definitions in order to achieve a particular emphasis in describing a particular machine or factory system. One fundamental common denominator is the removal of human labour from identifiable tasks which remain similar in form but are accomplished by machines. A more subtle and specific theme lies behind those definitions which rest on feedback and control: namely that this represents the mechanisation of mental labour (Cooley, 1980). This theme seems particularly relevant to discussions of the effects of 'automation' on workers. It raises issues such as the turning of skilled workers into machine-minders, the possibility of new skills associated with the machinery and the manner of their distribution amongst the work-force, and the wage rates associated with the change of working conditions.

Job Loss

The main stimulus to the discussion of the 1950s and 1960s was a heightened awareness of the labour displacement consequences of technological change. One clear reason for this, at least in the USA, was that the economic growth of the post-war period was over-shadowed by a significant but, as it turned out, temporary increase in unemployment in the late 1950s. Thus one finds that many of the contributions to the debate which focused on the displacement consequences of 'automation' came from trade unions. Moreover, the issue was seen as sufficiently substantial, or at least sufficiently sensitive, to warrant the establishment of a US Congressional Committee of Enquiry. Much of the literature of the time consists of submissions to this committee (US National Commission, 1966).

Of course it could be argued, and was argued at the time, that a second stimulus to this discussion was a genuine acceleration in the pace of technological change which was taking the form of a widespread diffusion of automated techniques into industry. The two reasons taken together amount to the now-familiar hypothesis of technological or structural unemployment. In fact, however, as Cohen (1975) notes, there was very little attempt to examine directly the rate of diffusion of 'automated' techniques into industry, perhaps because of the failure to define them sufficiently accurately to operationalise their definitions. Furthermore, those attempts which were made were quite simple and insubstantial.

Economic Growth

The discussion of the effects of 'automation' had a certain degree of overlap with a broader discussion which was developing at that time on the contribution of 'technical progress' to economic growth. Several observations have been made on the stimuli to this debate (see in particular, Nelson, 1959). Undoubtedly, the increased government involvement in research and development ('R&D'), the huge increase in military spending during the Cold War, the space programmes, the nuclear programmes and the prominence of new industries which were based on technologically intensive activities, all played a part. Despite the ever-increasing complexity of the work which followed from Solow's (1957) paper on the substantial contribution of technical change to economic growth, the points of overlap between this work and the debate on 'automation' remain limited. What did percolate into the public arena was the insight that technical change, even 'automation', might be amongst those factors which promoted industrial growth and therefore gave rise to new employment through new capital formation. The fact that economists remained unsure and even divided over exactly how this mechanism worked, whether it worked efficiently, and whether it might work differently at different times, prevented general awareness moving beyond this simple observation of the potential of technical change both to create and to destroy jobs. As one might expect, speakers from industry tended to emphasise growth and job creation, while trade unions expressed fears of unemployment and job deskilling.

We can summarise this period in discussion of 'automation' by saying that there was little progress in clarifying the nature of 'automation' (a notable exception being the work of James Bright (1958) discussed in the next section) limited progress in establishing the economic parameters of 'automation', both at the micro- and macro-economic levels; and little progress in producing any empirical data on the penetration of 'automation' into particular industries, let alone the effects of such penetration.

Later Discussions: Bright and Bell

We are currently in the second of the two periods of extended debate about 'automation'. Since the slowdown in growth of the developed capitalist economies which began in the early 1970s, increased rates of unemployment have revived fears of the adverse effects of technological change on employment. The notion of structural unemployment has become more refined and established and is backed up by more sophisticated analyses of the roles of capital and labour in production and by case studies of 'automation' in particular industries. A representative example of this type of analysis is Rothwell and Zegveld (1979). However, many of the same conceptual and practical difficulties which beset earlier discussions are still in evidence. There have been few notable additions to the literature on the definition of 'automation'. Therefore, the good reasons for the existence of the current interest in 'automation', namely reduced growth and increased unemployment, are further reasons for pursuing the task of clarifying the nature of 'automation' and examining the real extent of its diffusion.

The enormous range and variability of the actual technologies found in industry usually undermine any attempt at watertight definitions of 'automation', so much so that one of the most rigorous of analysts, James Bright, was forced to conclude in the mid-1950s that 'automation most definitely means different things in different companies and industries' (Bright, 1958, p. 52). Furthermore, he felt that the only accepted general definition of 'automation' was 'something significantly more automatic than previously existed in that plant industry or location' (Bright, 1958, p. 67). Behind this deceptively simple view, which apparently abdicates from attempting an analytical approach to automation, is the basis of one of the most complex and influential approaches to the subject yet produced. Bright deduces from his description of 'automation' as an evolutionary process that there must be a hierarchy of levels of 'automation', and that the evolution of 'automation' of a particular system can be seen as progression within this hierarchy of levels. These 'levels of mechanisation', as Bright calls them, are illustrated in Figure 2.1.

It can be seen that from level 4 onwards the main concern of the scheme is with the nature of *control,* the source of the initiating signal and the character of the machine response. Successive levels increase in sophistication, with the ultimate level (17) coming close to contemporary engineers' notions of artificial intelligence. It seems likely that the scheme can be applied equally easily to individual pieces of machinery or to systems of machinery. It therefore suggests that definitions of 'automation' based on transfer of work-pieces are incomplete, since transfer is often subsumed as a function within the levels of Bright's scheme. Bright's work confirms that what really distinguishes 'automation' from previous developments in mechanisation is advances in *control.* It should also be noted at this point that the analysis is not dependent on any specific technology. The means of control are not specified, only the degree. As we shall see, this approach has strengths and weaknesses.

Despite being formulated quite early in the supposed period of acceleration in the development of 'automation', Bright's approach to categorisation proved so influential that there were no serious rivals to it for nearly a decade and a half. It seems, however, that there was little enthusiasm, or perhaps little perceived need, to use Bright's work to interpret the history of technical change and chart the development of 'automation'. People who showed the most interest in this piece of applied management science were, paradoxically, Marxists such as Braverman (1974) and Mandel (1975) (and then not until the 1970s) who noted Bright's dispassionate conclusions concerning the reduction of skill levels implied at the higher levels of his 'automation' hierarchy. It will be remembered from our earlier discussion that in the 1970s a revival of the 'automation

26

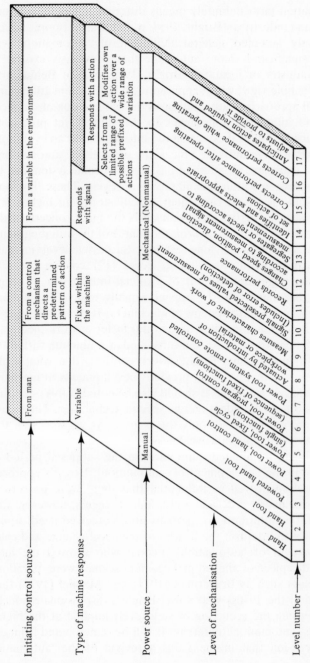

FIGURE 2.1 *Levels of mechanisation and their relationships to power and control systems*

SOURCE *Bell (1972) adapted from Bright (1958)*

debate' ensued and has continued more intensely in recent years. Although not integrally linked to this debate, it is at this time that we find a new and significant analytical contribution to the taxonomy of 'automation'.

Bell's (1972) work for the (British) Engineering Industry Training Board on forecasting the manpower implications of technological change involved him in a painstaking survey of technical trends in all branches of this industry, but by far the most important and developed part of this work concerns the mechanisation of metal-cutting processes. In studying the developments in this area, which have been significant since Bright's time, Bell found it possible to advance a more refined conceptual description of 'automation'. Although he does not claim any applicability for this scheme beyond his chosen area of engineering, we shall see that there are strong arguments for regarding it as a useful basis for a general purpose definition of the scope of 'automation', given certain objectives and conditions.

The main point in Bell's analysis is this. He disagrees with Bright that control is the only focus of 'automation', and argues that every manufacturing activity contains some elements of three distinct processes: *transformation* of the materials of workpieces into new shapes, states, assemblies or forms; *transfer* of materials or work-pieces from one part of a production system to another; and, third, *control* of the other two activities. (See Figure 2.2.) His basic insight is that the levels of mechanisation or, as Bell misleadingly calls them, levels of automation of these three parts of any production system

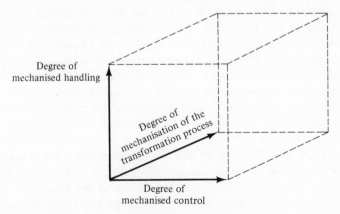

FIGURE 2.2 *Mechanisation as a three-dimensional space*

can be quite different. A particular machine or complex of machines can combine quite high levels of what Bell calls 'automation of transfer' with low levels of 'automation of control' and vice versa. The 'automation of transformation' is, however, usually quite high in all concrete examples of contemporary machinery in engineering.

In order to define a 'level of automation' (in Bell's terms) we need to modify Bright's scheme, where a level is simply a point on a spectrum. Instead, we need to speak of a point in a 3-D space, where the three axes of the space are the degree of 'automation' of transformation, of transfer and of control. This space can now serve two useful purposes. First, we can identify particular points in this space which correspond to real pieces of technology – Bell does this very convincingly for machine tools (and we refer to this in more detail in Chapter 5). Second, we can conceive of the historical process of technological change, which defines 'automation' as the movement of a surface within this space. The direction of change will vary with time and with the nature of the various pressures and constraints influencing each incremental step in the process. In fact, since machinery is never designed to one blueprint, but rather different machines for different purposes occupy different positions in the vector space, we should speak of the gradual movement of a surface rather than a single point defining technical possibilities. The shape of the surface is the determinant of the potential for 'automation'.

The debate over the definition of 'automation' is a complex one but it seems to us also to be an unnecessary one. As we will illustrate in later chapters, the various participants in the ' "automation"-defining' game have been trying to capture two different processes in one definition. Since World War Two the use of machinery has been advancing on two fronts. Incremental changes in *transfer* (using old and new technologies) and the diffusion of radical new *control* technologies have coexisted. Writers have struggled to capture these two different trends in some all-embracing notion of 'automation'. Consequently, some of them have emphasised 'feedback', others the grander systemic properties of large-scale integrated production processes, others the notion of mechanised assembly lines for particular tasks. In our view, the notion of 'automation' is therefore more of an obstacle than an aid to understanding what is actually happening in the development of production processes. It is more useful to differentiate between three *types of mechanisation* than types or levels of 'automation'. We propose, therefore, to use the terms, *primary, secondary* and *tertiary mechanisation* and to drop the term 'automation' as an analytical category. The three types refer to the three dimensions of mechanisation (which, confusingly, Bell calls dimensions of 'automation') already discussed; namely, the mechanisation of transformation (primary), of transfer (secondary) and of control (tertiary). We will use them to discuss the history of production systems in Chapter 3.

'Automation' and the Labour Process

The three-dimensional scheme which Bell designed for his examination of metal working, suitably modified in the way that we have described, has a generality and abstract quality which seems to make it quite serviceable for all manufacturing (and even some non-manufacturing) activities. Yet as Bell shows, it is capable of being made quite precise when required. He uses it to assess how the addition of new points to the mechanisation surface will affect the requirements for various types of skilled labour in the engineering industry. However, the scheme is not able to do this by itself since it does not include a crucial variable. When new technologies become available and are installed, a management decision has to be made concerning the job description of those people working with the equipment. There are always a large number of production tasks to be done and they have to be 'parcelled' into packages to make jobs. The skill level and the number of people required for each job depends on how the tasks are combined. While there may be a strong influence of the machinery to suggest a 'natural pattern' of jobs, there is nevertheless a great deal of scope for variation. This could be contested by one or more of the groups of people likely to be concerned with the outcome, amongst the management and the work-force.

For the present, we can note one very significant conclusion from this discussion. Bell finds that his forecasts of changes in skill requirements (both quality and number) suffer from a real indeterminacy because of the *negotiable* character, though within limits, of the combination of tasks into jobs. For our discussion this means that the development of mechanisation is not simply the replacement of one machine by another, more highly mechanised or more 'automatic' machine. Rather, it is the replacement of one *human-machine combination* by another human-machine combination. In other words the process of mechanisation can only be understood as change in labour processes. The technological aspect of this change is best seen as one component part of the process and perhaps, in most circumstances, contingent upon the other parts of the process. This is in striking contrast to the popular technological view which sees the arrival of the new machine as exogenous 'technical change' and the work organisation changes as merely contingent on that arrival.

In our view, therefore, there is a clear link between the technical literature on machinery and 'automation' and Marxist analyses of machinery as a semi-contingent part of the labour process. As we have mentioned in Chapter 1, it is only in the past decade that the implications of this aspect of Marxist thought for contemporary technological change have begun to be examined. One observation from labour-process analyses is particularly relevant to our attempt to identify more closely the nature of mechanisation developments. During the historical development of the capitalist mode of production, there has been a tendency for 'capital', through its management agents, to increase the degree of control over the labour process. At this point we must differentiate control of the labour process with which we are concerned here from the narrower, but related, notion of control of the actual manipulation of tools, etc., as described in the more technical literature of Bright and Bell. Managers are concerned with control in the broad sense. They take steps progressively to remove uncertainty and therefore discretion and skill from the labour process by modifying it, to bring the planning and conceptual activities under their direct responsibility and by constructing labour processes which predetermine the actions of workers more precisely. Braverman's extended critique of Taylorism as a means of labour control, and of the essential continuity with Taylorism of supposedly 'humane' work design approaches is the principal reference point of this mode of analysis. We will discuss in Chapter 3 recent challenges to this 'deskilling' dynamic.

There is a major ambiguity in the literature on the labour process. Some authors have tended to adopt the view that 'capital' requires control over the labour process almost for its own sake, that it simply reflects a continuing effort to maintain the wage relation which is the central pillar of the capitalist mode of production. Some, on the other hand, have seen control of the labour process as an objective which flows logically from the purely economic pressures on individual capitalists. Marxist economic analysis views competition between capitals as being likely, other things being equal, to promote technical changes which reduce production costs, increase capital intensity, reduce living labour input per unit of output and effectively embody more and more of the co-ordinating and planning function in the physical structure of the production process. This view sees 'valorisation' of capital as a more powerful objective behind capital's struggle for control of the labour process, in contrast to the more narrowly political or power-centred interpretation of the other group of authors. We shall return to this point in Chapter 5. At first sight, it seems likely that both these factors are important, but the problem is to describe exactly in what way they are linked.

So, what is the connection between this narrow tradition of labour-process analysis and the general (and non-Marxist) literature on 'automation'? The answer is that they provide a similar base for describing what 'automation' is and why it occurs, although they arrive at it through different routes of analysis. Marxists start from the notions of labour process and capitalist relations of production and deduce that continuing mechanisation of control will be a central aspect of the technological change which is used to make the labour process conform with the social relations. Non-Marxists, on the other hand, start from the observed characteristics of the machines and machine systems and find the changes in those characteristics cannot be deduced solely from 'natural improvements' or from exogenous technical advance, but rather from a complex of variables which look, upon examination, very similar to those which Marxists are using in describing attempts to modify the labour process.

Conclusion

We have, then, broken away from the early attempts to define 'automation'. They were inadequate, we would argue, because they all sought to define 'automation' by reference to technological or engineering criteria. The labour-process style of analysing 'automation', however, whether done in the tradition of Bell or of Braverman or of both, as in the case of the Brighton Labour Process Group (1977), offers a framework intrinsically more useful to the student of technological change. This is because of its ability to conceive of technology in a non-determinist way, as the result of a process of innovation which is competitive between different firms and between managements and workers in each industry and firm, and is also influenced by the general socio-political circumstances in which it occurs. Thus we can hope to gain insights into the relationship between innovation, types of mechanisation and capitalist economic development.

Unfortunately, possession of an acceptable starting-point does not guarantee arrival at useful conclusions. The problem facing all abstract frameworks is their practical application. In this case the task is to examine the actual history of mechanisation in terms of the parameters surrounding the labour processes and economic and social circumstances in which machinery was designed and used. Given that we acknowledge some relative independence of the evolution of the relevant knowledge base (that is, scientific and technological knowledge) as an extra complicating factor, this will not be a straightforward task. Despite the endeavours of historians of technology, knowledge of the vast range of industrial machinery and its use, and especially of the machinery of the period since World War Two, is limited. In this discussion we cannot hope seriously to expand the detail of that knowledge, but rather to contribute to the process of bringing some order to the wealth and diversity of detail in existing material. Our task is to identify more clearly how the different types of mechanisation have developed historically, what technological means are used to achieve mechanisation and whether these instances of mechanisation have any specific relation to the economic, social and political relationships in capitalist societies.

3 Mechanisation and the Labour Process from 1850

INTRODUCTION

Looking back over the past 130 years we can see some patterns in the development of production machinery. The patterns have expressed themselves in leading industrial sectors at different times since the mid-nineteenth century. In this chapter we shall try to identify how those machine systems measure up against the classification scheme we have developed from the work of Bright and Bell. Of particular interest, of course, are changes in the labour process which link machinery developments to the broader structural changes within the development of the capitalist mode of production.

It should be made clear at the outset that we shall not deal in detail with the wide variation in the rhythms of development across countries over the hundred-year span of this chapter. There were of course substantial differences between British and mainland European developments, between Europe and America and between the 'West' and Japan. For example, Littler (1982) has pointed out substantial differences between the USA, Europe and Japan regarding the techniques of management which go under the name of Taylorism and has shown that this particular 'ism' took substantially different forms in the three regions because of their differing cultures and of the results of class struggles over workplace organisation in the 1920s and 1930s. Nevertheless many of the basic techniques of Taylorism, transformed according to local cultures, were favoured practices amongst managers and engineers by the 1950s. So despite the different tempo and pattern of diffusion of Taylorist ideas, over a longer time-span their application can be seen as a particular pattern in the organisation of production in mid-twentieth century capitalist economies. We attempt to identify those long-term trends in technological and production organisation developments which, even if not always contemporaneously and in their original form, have displayed themselves in the major capitalist countries.

TECHNOLOGY AND THE LABOUR PROCESS: 1850–1900

Before addressing this topic it is necessary to outline the context, in particular the preceding developments in industrialisation. It is acknowledged by most economic and social historians that in the period 1750–1850 the gross transformations of the Industrial Revolution were completed in England – the creation of a national market, several manufacturing industries mature enough to export significant fractions of output, a considerable urban working class and so on. Though the changes from the 1850s onwards are, of course, immense, including the industrialisation of much of the north of the globe, they differ in one crucial sense from what had gone before. From this period onwards the changes in the industrial and economic system can be characterised as increasingly intensive changes in an already established mode of production, rather than predominantly the expansion of that mode of production from infancy to fully-fledged existence.[1] Of course, this judgement must be qualified by the fact that the capitalist mode of production is *still* extending its scope by penetrating new geographical areas and even new sectors of advanced economies.

　　What were the principal changes in production during this period, 1750–1850? Marx's description of changes in the labour process in *Capital*, Volume I, suggests the following, by now well-known, sequence of configurations: first, a transition in particular leading sectors from individual craft work with little division of labour to a phase of simple co-operation, involving several workers working together in one place; second, as the division of labour intensified, resulting in the fragmentation of the foregoing craft operations, detailed tasks and detailed labourers emerged. This created a number of individual labour processes which could utilise simple tools and mechanical aids which were more consistent with repetitive work than had been the craft tools which were their forerunners. This was the system which Marx called 'manufacture', meaning literally the making of things by hand by collectivities of workers in close physical proximity (that is, in what where first called 'manufactories'). At this stage, however, the labour processes of individual workers, and of the overall collective of workers, were for the most part still limited physically not only by the dexterity and sensory abilities of human labour, but also by its strength, even though this was often augmented in various ways. Thus, the next phase involved the lifting of this constraint by the systematic use of mechanical prime-movers. This was the period of increasing use of machine-made, steam-powered machinery which Marx characterised as 'machinofacture', by contrast with manufacture.

Machinofacture marked the beginnings of a transition of the capitalist mode of production from purely extensive expansion to a mature process which focused increasingly on intensive growth. Marx himself declared that it was only with the technical basis of machinofacture for the production of further machines that capitalism created for itself a 'fitting technical foundation' (Marx, 1976, p. 376). This transition to machinofacture began early in the nineteenth century in several sectors but only by the middle of the century did it become firmly dominant over manufacture in the construction of machinery.

Machinofacture was, it should be remembered, not simply a quantitative increase in productivity from capital's point of view. It was a distinct labour/production process which, as noted by the Brighton Labour Process Group (1977), marked the move from 'formal subordination' towards increasing 'real subordination' of labour by capital. These are the terms used by Marx to denote the distinction between the bare wage relationship (formal subordination) and its augmentation by *direct* control over the nature and pace of work via machinery (real subordination). This real subordination was a pre-condition for the pursuit of increased relative surplus-value and was central to the ability of the capitalist mode of production to sustain expanded reproduction of capital.[2] By the middle of the nineteenth century in Britain a production process of machinofacture was established in leading industrial sectors (particularly in textile production) and its evolution was intimately intertwined with the broad features of industrialisation. In the third quarter of the century the outstanding feature in the development of industrial production was the generalisation of steam-powered machinofacture throughout broad swathes of industry not only in Britain but in parts of mainland Europe and the USA. The data are well summarised by Landes (1969), who shows that the main growth in the use of steam engines in Britain and other European countries begins in the 1840s and that capacity doubled every decade until the 1870s. Moreover, this diffusion was not just through the firms of each industry but also within each firm; the steam engines were applied to a multiplicity of production tasks in the course of making a particular product.

What is the significance of this series of developments for analysis of the historical development of mechanisation? Landes characterises the period we have described as one of 'mechanisation', one of the several precursors to what he calls 'automation', which he sees as a late twentieth century affair (Landes, 1969, p. 323). From our conceptual discussion in Chapter 2 the trend of applying steam-powered machinofacture could be seen by contrast as a major, initial step in the *mechanisation of the transformation component* within the three dimensions of the labour process. This is more than a semantic distinction. It provides an interpretive framework which helps to understand the eventual constraints, as well as expansive properties, created by this phase of mechanisation. Subsequently these constraints contributed to shaping the production process 'strategies' of capitalists and the focusing of inventive and innovative effort toward initiatives which facilitated a second, quite different, phase of mechanisation.

It is necessary to introduce a qualification to the analysis at this point. During the period discussed above, the particular phase of mechanisation taking place was relatively evenly distributed across several industrial sectors. It did not of course reach all sectors, nor all operations in any one sector, but it was sufficiently general to enable us to speak of it as a dominant mode of technical change. However, as the analysis is pursued further toward the present day, this uniformity breaks down. The phases of mechanisation which are identified below occur at different times, to different extents, and occupy different lengths of time in different industries. It is not possible nor appropriate in this account to document all this heterogeneity since the concern is to explore some generally applicable theoretical ideas. Henceforth, therefore, the object of analysis is a generalised production process, mirroring the leading edge of capitalist industrial development. At some points it will more nearly approximate to one industry, and at some points more nearly to another. Where possible these specific connections will be made explicit. This method is, it is hoped, valid for the purpose at hand, namely identifying the major types of mechanisation which have developed, what factors have influenced them and what their effects have been.

A SECOND PHASE OF MECHANISATION

In the period from the last decades of the nineteenth century to the end of the first war the pace of technical change, both innovation and diffusion, was extremely rapid.[3] The most prominent change was the arrival of new sources of power to replace the steam engine. These were of course the internal combustion engine and electricity. The series of incremental innovations which had greatly augmented the efficiency of the steam engine had seemingly diminished and the physical constraints imposed by steam power were increasingly becoming a commercial problem in the way that the limitations of human strength had been before them.

The internal combustion engine was the first radical innovation to break this log-jam. Though it did experience a significant period of importance as an industrial prime-mover it was not to begin a really massive diffusion phase until the coming of the automobile, which is a later phase in the story. Instead, it was overtaken by electric energy in the field of industrial power. The reason for this probably lies in the relative costs of their respective primary energy sources at the time. Petroleum was still an under-developed resource, whereas coal was a mature product with a well-developed distribution system. The steam-driven turbine driving an electric generator possessed the advantages of greater primary efficiency than reciprocating engines in extracting energy from heat. It also had the ability to provide a very mobile, flexible and reliable flow of energy which could be used in variable quantities when, where and how the user required. Against this should be set, as a great barrier to its diffusion, the considerable difficulties involved in creating a distribution system for electricity. However, it can also be argued that it was precisely in this distribution system, with its consequent guarantee of an eventual market amongst domestic consumers as well as factories, that the great incentive for adoption lay. The great significance of electricity for the production process and for further mechanisation lay in its gift to the factory owner of mobility of power sources, though this potential was not fully realised until some time later. Other aspects of change in the production process at this time also deserve mention in order to assemble a more complete picture.

The mechanisation of *transformation*, given such impetus in the previous period, was characterised principally at this time by the achievement of faster and faster speeds of operation of the transforming equipment. This rested on a wealth of technical innovations, some minor, some fundamental, many of them consciously sought solutions to perceived problems in production. Landes mentions the following; high grade steels for tools which improved cutting speeds, making the machines themselves grow in size in order to provide these speeds; faster-operating textile machinery; the improvement of lubrication techniques and the use of ball- and roller-bearings. The replacement of hammering by rolling in steel and wrought iron production was particularly noteworthy since it initiated the transformation of that industry into the first of what we now call the continuous-flow production industries.

But this great increase in the speed and scale of transformation, considerably stimulated by foreign as well as domestic competition, brought with it new problems. The great power, size, speed and expense of the transforming equipment merely served to underline the inadequacy of the way in which work was organised to flow *between* operations and the residual variability in the manner in which it might be performed. To quote Landes:

> The entrepreneurs of the late nineteenth century were thus goaded by necessity and spurred by the prospect of higher returns to find ways, first, to ease the movement of work through the plant, and second, to draw more output from each man with a given body of equipment. (Landes, 1969, p. 302)

Yet it would be necessary to do more than 'speed up' workers; the nature of the individual jobs would, of necessity, have to change to accommodate the new organisational patterns associated with revised work-flows.

This suggests that the progress of mechanisation and technical change was neither random nor pre-determined. It proceeds at least in part via a succession of particular foci, which can be usefully captured in the image of the 'bottleneck'. At any point in the development of a production process, given a particular structure of competitive pressures, there will be certain nodal points where the penalties, the potential rewards to change and the means to achieve change, will form a volatile combination. Effort will be focused on this nodal point and technical and organisational change will be biassed toward relieving the bottleneck. This general phenomenon is described in more detail by Rosenberg (1976), in his essay on 'inducement mechanisms and focusing devices'. Two of his categories of inducement are particularly relevant here: first, the presence of imbalances between interdependent production processes and, second, groups of workers exercising (from management's point of view) awkward degrees of control over parts of a production process. Both these phenomena were prominent causes of quite deliberately sought innovations throughout the nineteenth century. What is suggested here is that, towards the end of the century, these mechanisms shifted the emphasis of mechanisation away from simple transformation problems and towards both transformation *and transfer* problems, thus bringing into play the second of the three dimensions of mechanisation which we are investigating. It is appropriate now to provide some more details of the way in which this change occurred.

The moves to improve the logistics of the flow of production were of two broad types. One fairly simple trend was most visible in those industries which produced large quantities of a relatively homogeneous product: textiles, chemicals, metallurgy, glass, petroleum-refining and basic foodstuff-processing; in short, the 'flow' industries. Another, more complex, trend was visible in industries which involved some element of assembly or fabrication. The technical problems of moving liquids, solids and gases at all temperatures gave rise to all manner of conveyors, elevators, hoists, valves, pumps, tanks, meters, gauges and controls. In several of these technologies, particularly those concerned with large-scale handling of solids, electric power sources with their previously mentioned mobility and rapidity of action were particularly appropriate. Thus, electricity was involved in both exacerbating the problems by speeding up transformation, thus moving the bottleneck to the transfer system, and in solving them by providing the means to mechanise transfer to some degree.

The dominant sector of the second group of industries, those which assemble or fabricate, was engineering. The mass-produced consumer durable industries were in their infancy. There were a number of inter-linked problems for capitalists in trying to improve transfer efficiency in these industries. First, work was often organised spatially according to machine type – drilling, turning, etc. This meant that work-pieces had to pursue a tortuous zig-zag pattern back and forth between sections of the plant before reaching completion, thus spending too long 'idle', away from the machine. It also meant that the organisation created centralised groups of craft workers who were in a position to maximise the bargaining power implicit in their skill and their ability to disrupt production far beyond the boundaries of their own specific tasks. The second problem, while appearing to be more narrowly technical, also related to craft skill. The materials, cutting techniques and measuring techniques of the nineteenth century did not, in the main, allow assembly to be conducted with randomly chosen parts, from stock. The precision implied by *interchangeability* of parts was beyond the majority of industries and enterprises. Not only did this prohibit what we now regard as normal, namely the highly productive assembly-line, it actually made assembly the skilled domain of the craftworker, who was alone capable of 'fitting' one part with another.[4]

The technical contributions to solving these problems were several. The increasing use of steel instead of wrought iron allowed greater accuracy, and the use of gauges allowed knowledge of accuracy. The milling machine, especially when coupled with improved tool steels, permitted accuracy in more complex-shaped parts. Precision grinding, both of parts and of tools, permitted the fine manipulation of dimensions still beyond the range of the machine tools themselves. The turret lathe, which carried a number of differently shaped tools, and which was later capable of rotating the turret and inserting and feeding the work automatically, was a particularly important innovation. It combined an increase in the mechanisation of transformation with some degree of mechanised transfer. As we shall see later it also contains a small step on the ladder of mechanised control.

The organisational contributions which attempted to solve the logistic problem were no less radical and impressive. The most systematic of these was Taylorism. Either in the form devised by Frederick Taylor by 1910 or in techniques derived from Taylor in the subsequent twenty years, for example by Bedaux (Littler, 1982) Taylorism sought to prescribe work tasks. This often entailed the redesign of those tasks on the basis of maximum standardisation. With the establishment of planning departments to organise and monitor shop-floor work, actual production tasks would become, in effect, routinised (or in Littler's word 'bureaucratised') and workers would be expected to have 'minimum interaction' with their work.

Taylorism, as we are using the term, involved attempting to treat the human actions in the production process as subject to the same principles of mechanical optimisation as the inanimate functions. Such principles were of course inconsistent with the worker retaining *sole* knowledge of, or *complete* discretion in, the choice of the physical and mental manipulations associated with the task. This process is, in one sense, nothing more novel than the continued application of the principle of the division of labour which is at the root of the development of collective labour processes from co-operation, through manufacture and machinofacture to the present day. But Taylorism had other features.

In the first place – and in the short term – it contributed to the changes from the turn of the century onwards which helped restore the trend of increasing productivity. It sought to reduce some of the uncertainties and the 'porosity' of the working day which managements needed to control. It was a major extension of real subordination of the labour process and, as Aglietta (1979) notes, it was a very appropriate response to a situation in which the production process was not *integrated* by the technology itself. It is worth noting however, that Taylorism was easiest to apply in new industries with little previous history of craft working. When meshed with the handling and transfer technologies developing throughout industry, in the context of large consumer markets, it became redefined within the classic 'flow-line' to which we shall turn shortly. In the second place, we should note that Taylorism, with its emphasis on *control* in the broad sense of authority, *and* in the narrower sense of control of manipulations laid the first step in the later process of mechanisation of control. Both these themes, mass-production and mechanised control, form the subject matter of later sections of this chapter.

Let us recapitulate. The commercial and technical pressures which existed at the end of the first phase of machinofacture shifted the balance of the process of mechanisation so that it began to advance on two fronts, transformation and transfer. The key enabling steps in this were new and more suitable power sources, innovations in handling devices, interchangeability and precision manufacture, and Taylorist forms of work organisation. All this, however, must also be seen in the context of the gradual changes on the demand side of the economy. The most mature expressions of these tendencies were found in the nascent high-volume consumer-good industries making such things as automobiles, sewing machines and bicycles and the industries which supplied them with parts or materials. Therefore the real maturity of this potent combination of techniques was not achieved until after World War One.

BETWEEN THE WARS: FORDISM

Although the inter-war period brought many significant innovations in mechanisation and in the organisation of production processes, the dominant aspect of this period must be judged to be consolidation of the changes which had had their genesis and had accumulated in the previous period. We have outlined above what these changes were, and Landes characterises the inter-war period as one of 'working out' these innovations (Landes, 1969, p. 41). This can be amplified by saying that there was extensive diffusion of flow and assembly line techniques, and intensive improvement of them achieved by incremental innovations.

It can be argued, however, that what has been characterised as a quantitative change had in fact an important qualitative character. Aglietta (1979) proposes that the transition from Taylorism before World War One to fully-fledged 'assembly line' production in inter-war America of many consumer products represents the establishment of a new production-process type with a significance beyond the point of production. This judgement is part of a complex and wide-ranging thesis concerning the dynamics of capitalist accumulation which cannot be fully discussed here (though we return to it in Chapter 7) but certain of the key features will be outlined since they constitute a useful contribution to the analysis of the development of mechanisation.

Aglietta labels assembly-line production as 'Fordism'. He does not rigorously define it, but from his and others' work it can be more precisely characterised.[5] Fordism describes in the first place a particular configuration of the production process associated historically with those industries which have provided the most visible examples of the consumption patterns of advanced capitalist countries, such as automobiles and household durables. Indeed, Fordist principles of production organisation make up the most dominant 'paradigm' for engineers and managers of large industrial and commercial units, though the term Fordism may not itself be used. The success of any putative new mass consumer industry depends on the extent to which Fordist principles can be applied to it; other industries where mass markets are unlikely to develop (for example in capital goods), measure the 'success' of any production organisation changes which are being contemplated by the extent to which they measure up to some notional idea of what could be achieved, in terms of productivity per worker for example, where Fordist principles to be economically applicable.

What constitutes a Fordist production process? Fordist principles combine various technological, organisational and labour management innovations in the production of a wide variety of goods. It is important to stress that the principles are applied to more than just the automobile industry where they were first put together by the Ford Motor Company. The principles are:

1. The production of large volumes of standardised products, using interchangeable parts.
2. A dedicated serial production process, often, but not always, facilitated by the mechanisation of transfer activities ranging from the low level mechanisation of the conveyor belt assembly line to the highly mechanised flow processes of food and chemical production.
3. A division of labour between 'deskilled', material-manipulating, machine-minding and machine-feeding workers (usually on the assembly line sector of a process) and skilled metalworking, maintenance, technical, planning and supervisory workers who are less machine-paced, if at all.
4. Amongst unskilled workers an acute division of labour, consonant with the employment of the historically cheap labour of women and migrant workers in these tasks. (Male dominated car assembly must not be thought of as typical of Fordist principles here, as Liff (1982) has pointed out.)
5. A set of techniques for the management of this labour; achieved by a combination of Taylorist and systematic management techniques such as the bureaucratic allocation of individual tasks, effort level payment systems, tight supervision and the application of work planning methods (for example, through time and motion study).

Fordism is, in short, a particular configuration of the labour/production process consisting of a sequentially integrated core–periphery system of production incorporating highly prescribed tasks and specialised production machinery, making very large batches of standard products at very high levels of productivity.

Given the importance which many labour-process writers, following Braverman, place on *Taylorism* as a defining characteristic of those production systems typical of twentieth century monopoly capitalism, we must explain why we prefer to emphasise Fordism instead, and indeed to subsume Taylorism under it. We have already noted that Braverman describes Taylorism as a set of techniques by which management seeks the monopoly of knowledge over how production processes (including workers' physical movements) should be organised so as to 'control each step of the labour process and its mode of execution' (1974: p. 119). Such an approach, argues Braverman, allows managements to attack craft skills. However there is much historical evidence that Taylor's techniques were most successful with *less skilled* tasks and had greater effects on factory *management* than on the control of workers (Nelson, 1975, p. 74; 1980, p. 140). Taylorism supplied a set of bureaucratic techniques to attack the power of foremen and internal sub-contractors ('gang bosses') rather than skilled workers, by creating a more direct employment relationship (Littler, 1982). Further, Taylorism provided means to develop standards for effort level setting and, to this end, increase the 'transparency' of the labour process by *sharing* knowledge previously monopolised by workers, rather than management itself monopolising that knowledge.

Even if management should wish to monopolise the knowledge associated with the production process, this might not be beneficial to them in pursuit of their long term goals. Managements must seek the co-operation of the workers for production to take place at all and this might conflict with Taylor's preferred techniques of bureaucratic task prescription, individualised payment systems and hostility to the existence of group working. Other management strategies which explicitly recognise the dependence of production on labour are possible; they allow degrees of worker 'autonomy' in exchange for commitment to broad management-decreed production goals (Friedman, 1977; Cressey and MacInnes, 1980; Littler and Salaman, 1984). Taylorism as an historical phenomenon was conceived in relation to specific problems of indirect forms of employment like subcontracting in turn of the century engineering but, as it turned out, some Taylorist techniques (for example, time and motion study) were particularly appropriate to the setting of effort levels and to making economies in workplace and task organisation in *Fordist* organised industries. The 'control' which such techniques permit depends on whether various tasks *can* be bureaucratically prescribed and whether or not there are better alternatives. Even then, Taylor's Scientific Management must still rely on some complicity on the part of the workers, encouraged by incentive payment schemes. After all, if Taylorist work routines were so good, then 'working to rule' would refer to a situation of strong managerial control, rather than to a form of collective worker resistance!

Taylorism, then, differs from Fordism in that the former describes a set of techniques for the management of labour and the latter a form

particular configuration of a production system seen as a combination of process and product technologies with human labour in a specific form of work organisation. The use of Taylorist labour management techniques in Fordist industries may be particularly efficacious under certain circumstances, but it is only one approach to labour management. For this reason then, it is not appropriate to use Taylorism as a central conceptual tool in trying to understand changes in the organisation of production processes, since this privileges the control of labour as the central concern of managements.

To return to our account of the origins of Fordism, it is usual to see the principles of Fordism first being put forward in the second decade of this century by engineers and managers of the Ford Motor Company, at least as a total package (Chandler, 1964). Individually none of the principles was new. For example, the technologies of sheet-metal stamping and electric resistance welding, techniques crucial for producing the parts on which the Fordist production of cars depended, emerged from innovations within the US bicycle industry of the late nineteenth century. In the case of the sewing-machine and bicycle industries, though firms in these sectors pioneered the civilian use of standardised interchangeable components in their products, they sold their products at the top end of the market, thus inherently limiting the market size to that created by the rising disposable income of the American middle classes. Ford on the other hand, aiming his production at a broader market, sought to overcome this limitation by dramatically cheapening the production of cars (Hounshall, 1981a).

There are several important implications here. First, Ford's success lay in his ability to reorganise the basis of production around the use of simple (unskilled) rather than complex (craft) labour. It was this which enabled the swift increase in volume to make possible the continued growth of the Ford company in the 1920s.

Second, Ford's success in achieving the economic conditions of mass production must be understood with respect to his unique position as a monopolist in the new car market which he created. During the ealy period of Ford's production of his famous Model T, it was the absence of competition that enabled him to pay his workers $5 a day, though the success of his methods meant that rival companies quickly followed his lead as the market for cars boomed in the 1920s (Chandler, 1964).

Third, in achieving a progressive cheapening of production by means of increased division of labour, standardisation of parts, routinisation of work and intensification of labour and the use of new technologies, Ford was able to maintain this position as market leader. However, such a position relied on continued growth and continual cost reductions; these turned out to be untenable. A combination of competition by product differentiation (developed particularly by General Motors, Ford's rival) the costs of model changeover, the coming of trade unions to car manufacture and the effects of the depression, ended the dynamic of the economy of scale upon which Ford's growth had depended.

Fourth, and most important, Ford sought to reorganise consumption as much as he reorganised production. He not only reduced the production costs of cars but also created a network of dealers and customer credit schemes to gear up consumption to mesh with the continuous high-volume production. It is this additional feature of the transformation of consumption under Fordism which Aglietta addresses.

Aglietta argues, as we have, that Fordist assembly-line production entails a radical increase in the mechanisation of handling and the fixing of workers to work-stations with the tasks organised according to Taylorist principles. He also argues however, that the productivity of such processes is so great that 'capital' can only generalise them *if* it takes active steps to guarantee some stability in the mass market necessary to absorb the great volumes of products. These steps involve the promotion of a working-class 'social consumption norm' which is qualitatively and quantitatively 'adapted' to the types of production being developed. Some of the components of this social consumption norm are major items like housing and cars which require new financial arrangements to allow their purchase (consumer credit, cheap bank loans, rental schemes, etc.) as well as the provision of a basic infrastructure of roads, electricity grids and so on. Other components are less expensive but quite potent in their influence on life-style and on the manner in which labour power is reconstituted away from work, for instance, labour-saving consumer durables. Together with the material elements of the consumption norm there are institutional changes such as the dominance of the nuclear family unit and the provision by the state of certain guarantees, through insurance, of a minimum level of consumption.

It will be clear from these brief comments that this theme in Aglietta's argument (to be discussed further in Chapter 7) of the link between a transformation in the production process and a transformation in the nature of consumption is a very ambitious one. He asserts that the establishment of this connection between production conditions and consumption conditions marks a new phase in the accumulation of capital. Its stabilising value to the accumulation process lies in large measure in the fact that it increases the possibility for the capital goods and consumer goods sectors of the economy to grow in a broadly balanced fashion. Aglietta sees this balance between the major departments of production as vital to the sustained post-war period of accumulation.

This link was not established either rapidly or painlessly, however. The depression of the inter-war period, with capital goods production in fact growing faster than consumer goods, with these consumer goods being restricted to luxury markets, is seen by Aglietta as a process of restructuring necessary to establish Fordism fully. It is certainly true that there was a dramatic increase in the sales of certain consumer durables in many countries in this period.[6] This is sometimes obscured by the images of the Slump of the 1930s, yet it does mean that the expansion of capacity in these industries, the economies of scale, and the newness of the investment made these industries centres of best-practice Fordist mechanised labour processes. These industries, in particular the car industry, had important effects on technical and labour process changes in upstream and downstream related industries. A particularly graphic example is the development of the continuous wide-strip mill for the production of sheet steel. It has been noted already that the earlier substitution of rolling for hammering has been a major step towards continuous operation. The technical potential of this step was fully exploited once it became clear that enormous demand for high-quality sheet steel was being created by the car industry and other consumer-durable industries. Once more, however, it is important to add the qualification that, in Britain in particular, the recession's effects were strong enough seriously to retard the diffusion of the continuous-strip mill.

Changes were also taking place in other industries. The traditional textile processes were relatively stable during this period. But, in complete contrast, there was an explosion of new products and techniques in synthetic fibres. With the exception of rayon, these products only grew to mass-production scale in the period after World War Two, but it is in the 1930s that the initial spurt of change must be placed (Freeman *et al.*, 1982). The production of synthetic fibres, and of the other new synthetic materials whose introduction straddles World War Two, gave a further stimulus to continuous-flow production techniques, though of course based on principles different from those used in the steel mill. These examples of a different type of transfer mechanisation are in fact closely connected to control mechanisation which is discussed later, but in acting as test-beds for more sophisticated systems of control, they were in fact the exceptions which proved the rule. The principal characteristic of the transfer mechanisation which became best-practice in the inter-war period was its rigidity and low level of control sophistication.

While transformation and transfer were achieved with less and less direct human involvement, control was established by one of two means. Either, control was achieved through continuous human monitoring and manipulation of powered devices, or else control in limited segments of the process was achieved via simple tailor-made devices in the machine itself. An obvious example of this latter technique is the system which uses the arrival of the work-piece to trip a switch to activate the transformation operation. The important feature of this sort of mechanisation, which is often called 'hard' or 'dedicated automation', is that it is designed to operate on only one specific piece of material and is useless, without redesign and layout change, on any other product. Its inflexibility therefore means that the cost of the equipment can only be recovered if a very long production run of the related part can be assured. Hence its appropriateness to a class of industries dedicated to the mass production of uniform commodities in quantities never before attained.

The only alternative to 'hard automation' was continued reliance on human skill. Indeed, even those processes which were 'hard automated' still required some human intervention and were often only part of the overall production process for the completed commodity anyway. But reliance on skill in control, especially in relation to large-scale production processes, brings with it new problems for management in ensuring increases in productivity. Admittedly, a car assembly line, to take an example, is subject to more managerial authority than a workshop of skilled turners; moreover, the technical level of control is also higher since the serial operations involved in assembly are a result of systematic pre-planning and machine-pacing rather than relying on the craft workers' judgements. In that sense, mechanisation of all three dimensions is higher. But a new sort of collective response on the part of the work-force is potentially present; and the interconnectedness and scale of the process renders it vulnerable to the smallest disturbance. There is no shortage of evidence from post-war plants with this sort of Fordist production process of the particular and dramatic nature of the conflicts which occur beween management and work-force.[7]

To this well-known limitation of Fordism, Aglietta has added some others. One is the fact that particularly intense examples of Fordist production produce mental and physical exhaustion amongst workers which cannot be 'recuperated away' even by the benefits of the social consumption norm. The dysfunctional character of this phenomenon for capital is its consequential absenteesim and sabotage, which adversely affect the productivity of plants. This is particularly the case where the labour market is tight and labour is collectively strongest. Yet many of the worst jobs have been filled by workers from the secondary, peripheral labour market (women, minorities, migrants). Another problem which Aglietta notes is the increasing difficulty of 'line-balancing' as cycle times for individual jobs change frequently either through trade-union efforts or as a result of partial technical changes. Although different in detail it can be readily seen that these problems are all closely related, and together they form components of a new 'bottle-neck' in the search for productivity and can therefore be seen as a potential new focus of technical and organisational change. Once again the partial solution of one set of problems has, in due course, transferred attention to the next weak link in the chain.

To summarise this section: the inter-war period was one in which the mechanisation of transfer (or secondary mechanisation) took a large step forward thanks to the combination of two factors. First was the maturity of the technical possibilities which had been assembled prior to that period. Second, we note the new circumstances of demand evolving in a cluster of related industries which made the application of secondary mechanisation particularly appropriate. There were however, signs of new developments in control (tertiary) mechanisation and of circumstances which might promote them further. Neither of these developments, however, took place to a great degree until the period after World War Two. It is to this period that we now turn.

POST-WORLD WAR TWO DEVELOPMENTS

Before embarking on an analysis of the developments in control mechanisation, it is important to emphasise the role of diffusion of transfer mechanisation in the long boom after World War Two. While some parts of industry in the USA had reached quite high levels of plant integration between the wars, this was in the main not quite so true of Europe. In the period after World War Two, however, many of the consumer durables which used these production techniques made the transition from luxury items to mass items owned by large percentages of the households in any advanced country. Furthermore, it is important to note the enormous stimulus to these techniques created during the war itself, by the demands for armaments under conditions of semi-command economic policy. Data on the diffusion of some typical transfer mechanisation systems can be found in Nabseth and Ray's (1974) classic study on diffusion. These show that between 1950 and 1968 the percentage of car-engine blocks machined on 'automatic transfer' lines in Britain rose from 21 per cent to 52 per cent.

It must surely be the case that this rapid diffusion of the latest forms of 'hard automation' played a considerable part in fuelling the popular view that so-called 'automation' was 'taking-off'. Important though this diffusion was, however, it was technically anterior to the developments in control mechanisation which were taking place at that time. Their diffusion was not at that time so rapid, but their impact on observers and on events was quite marked. Given that these two trends were occurring in parallel it is perhaps not surprising that the rash of definitions referred to in Chapter 2 straddled uneasily the ground from Detroit automation to feedback principles. As we have argued, the observers were trying to find a one-dimensional term for a two-dimensional phenomenon.

What then were these developments in control technology? The most researched is the numerical control (NC) of machine tools. The technical components of this type of machinery have their origins in the tight links forged between the machine-tool companies, the aircraft industry, and the Defense Department in the USA during World War Two. The rapid change in design and sophistication of aircraft stimulated by the war, and by post-war civil possibilities led to demand for precision parts to be manufactured cheaply, but with frequently modified specifications. This of course is just what dedicated secondary mechanisation *cannot* do. The solution, a system of controlling the path of the tool which could be reprogrammed, was therefore consciously sought from 1949 onwards with the help of a development contract between the US airforce and the Massachusetts Institute of Technology Servo-mechanisms laboratory (Atkinson, 1980; Noble, 1979). The system was introduced commercially in 1955 and its diffusion has been steady since then. An impression of scale can be gained from the fact that NC tools represented 20 per cent of the value of US machine tool shipments in the late 1960s and in the early 1980s they represented 50 per cent of the Japanese machine-tool industry's output.

In developing the theme of control mechanisation we now begin to complete the picture of the three-dimensional progress of mechanisation using Bell's framework. NC technology in its pure form is in fact only one point in a series of levels of control mechanisation and has in fact been surpassed in recent years. This spectrum is shown in Figures 3.1 and 3.2, from Bell.

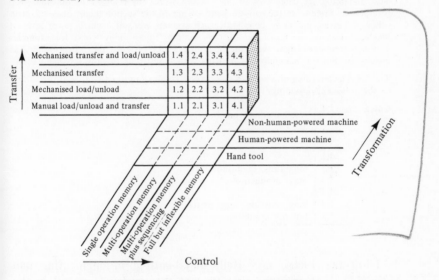

FIGURE 3.1   *Mechanisation of metal cutting in the 1960s*

SOURCE   *Bell (1972)*

The status of mechanisation in the early 1960s
The various forms and levels of mechanised cutting equipment available to industry in the early 1960s can be placed within the framework in the figure. In summary the three dimensions are:

1. *Basic transforming systems*
   Highly mechanised with most machine tools at this level.
2. *Control Systems*
   Four categories of mechanised system, distinguished largely in terms of input-memory capacity.

3. *Handling systems*
   Four categories of systems ranging from non-mechanised to highly mechanised transfer and loading systems

This appears as a two-dimensional section of a three-dimensional mechanisation space in the figure. Only about ten of the sixteen possibilities were practical realities:

1.1 Conventional, manually controlled machine tools
2.1 Capstan and turret type machine tools
3.1 Tracing and copying machines, manually loaded 'automatics'
4.1 Manually loaded special purpose machines
3.2 Mechanically loaded 'automatics'
4.2 Mechanically loaded special purpose tools
3.3 'Automatics' with transfer systems but manual loading–unloading (an unlikely combination)
3.4 'Automatic' link lines transferring similar parts between 'automatic' machine tools
4.4 Transfer lines.

The remaining six categories of the section were not relevant to machine-shop production. Systems in the top-left hand corner of the section might be relevant in other processes, but more commonly such systems were at a much lower level of mechanisation of the basic transforming system. In assembly work, for example, although frequently there were mechanised transfer systems, the transforming was usually carried out manually, or perhaps with hand tools or powered hand tools.

The ten relevant categories are closely related to scale of production roughly according to the following groupings:

Categories: 1.1 ⎱ small–medium batches
            2.1 ⎰

            3.1 ⎱ large batches
            3.2 ⎰

            4.1 ⎱
            4.2 ⎱ long runs, continuous
            3.4 ⎰ mass production
            4.4 ⎰

            3.3 ⎱ long runs and large batches
            4.3 ⎰ but not common.

*   *   *

Until the 1960s, says Bell, metal-cutting in engineering had available to it only the first four of the levels of control in the diagram, in combination with four possible levels of transfer. Combination 4.4, the most advanced, was the transfer line referred to earlier as 'hard automation' (See Figure 3.1.) Level 5 (shown in Figure 3.2) is a forerunner of NC known as 'plug-board information input'; level 6 is NC itself; level 7 is a development of NC known as Direct Numerical Control (DNC) in which the making of a tape for each machine is obviated; level 8 is adaptive control, in which feedback systems on a wide variety of parameters give the system a 'real-time' flexibility of response to add to its flexible memory. It will readily be seen that if these high levels of control mechanisation are

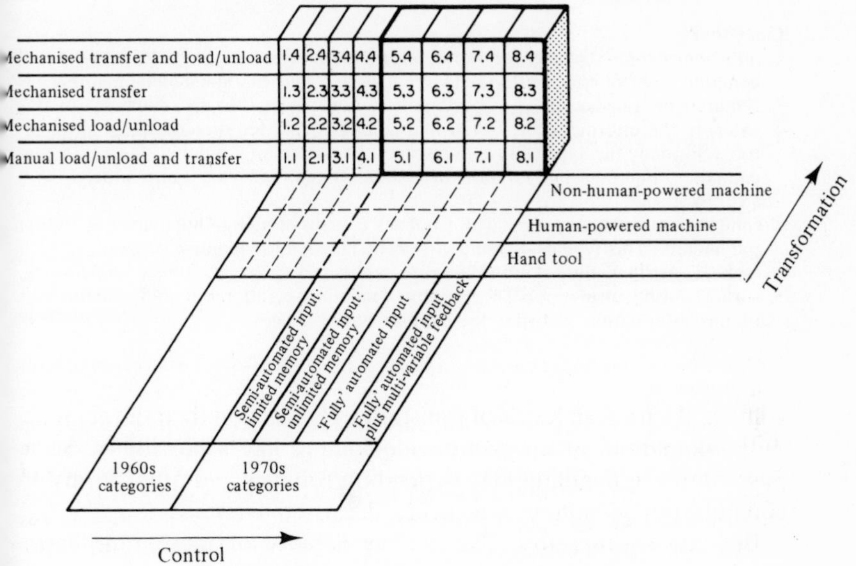
Mechanised transfer and load/unload
Mechanised transfer
Mechanised load/unload
Manual load/unload and transfer
1.4 2.4 3.4 4.4 5.4 6.4 7.4 8.4
1.3 2.3 3.3 4.3 5.3 6.3 7.3 8.3
1.2 2.2 3.2 4.2 5.2 6.2 7.2 8.2
1.1 2.1 3.1 4.1 5.1 6.1 7.1 8.1
Non-human-powered machine
Human-powered machine
Hand tool
Transformation
Semi-automated input: limited memory
Semi-automated input: unlimited memory
'Fully' automated input
'Fully' automated input plus multi-variable feedback
1960s categories
1970s categories
Control

FIGURE 3.2 *Mechanisation of metal cutting in the 1970s*

SOURCE *Bell (1972)*

Machine shop mechanisation in the 1970s

In summary, the developments in the 1970s can be described in terms of the basic framework thus:

1. *Basic transforming systems*
   Similar to those of the 1960s.
2. *The handling system*
   Wider and more flexible applications of the three categories of mechanised systems available in the 1960s.
3. *The control system*
   Four new categories of control system, thus:

Category 5:
'semi-automated' transfer of input information, with limited memory. This category describes the plug-board control, program-sequence control and possibly camless-auto types of system. In all of these, most of the information for a full cycle of operations can be prepared as plug locations, punched cards, etc. A machine can be rapidly set up with the information and left to 'read' this and operate accordingly. Usually only a limited number of possible variations in the information can be held in these 'pre-packaged memories'.

Category 6:
'semi-automatic' transfer of input information with unlimited memory. This category describes numerical control systems, which are very similar to those in category 5. Normally most of the necessary operating information for the full cycle of a job is coded on paper or magnetic tape. This is manually loaded to a tape-reader machine control system. A change of jobs requires, normally, a manual change of tape together with any necessary tool setting. The information capacity of the tape memory is almost unlimited, but the machine tool itself has limited capabilities.

Category 7:
fully mechanised transfer of input information. This category describes direct computer-control systems. In the two previous categories some manual assistance is required to transfer the pre-coded information package to the machine. In this category the information is stored in a computer and fed directly to the machine tool. Normally this type of system involves the control of a number of tools by one computer, but the configuration of control system can vary quite widely.
Category 8:
computer-controlled systems with feedback control of a large number of operating parameters. This type of system tends to be labelled as 'adaptive control'. In such systems, various output or operating parameters (e.g. workpiece dimensions, surface finish, cutting speed) are measured and the operation is modified in the light of this information in order to optimise performance.

         *   *   *

combined with high levels of transfer mechanisation then the genuine 'full automation' of the pundits looks more like a possibility. Such possibilities in handling may depend on robotics and other forms of computerised handling, which are discussed later.

Bell argues therefore that the availability of these four extra notches of control mechanisation since the 1960s marks a real discontinuity in the historical development of machine-shop mechanisation. This discontinuity lies in the fact that they render small-batch metal-working production potentially amenable to the same levels of mechanisation and unit-cost reduction formerly restricted to mass production and 'hard automation'. Since perhaps 80 per cent of production in engineering consists of small batches of less than 1000 articles, the significance of even the partial tertiary mechanisation of this small-batch production can easily be appreciated. Furthermore, NC machines, either singly or in combination, can be used at key points in more traditional mass-production systems, and can even create new types of mass-production systems. They begin to blur the old line between mass- and batch-production. Writing in the early 1970s, Bell was cautious about how quickly these systems would be introduced and what their effects on employment patterns would be. He believed that these technologies would break the previous relationships between technical change and employment, removing some unskilled labour from residual transfer tasks and de-skilling some control tasks. This suggests that there may be a considerable production-process change involved.

This is indeed the view of Aglietta. He appears not to be familiar with Bell's work but his diagnosis of the significance of this particular type of mechanisation is essentially similar. He labels it 'Neo-Fordism', following Palloix (1976), and gives it two principal characteristics. First, there is the technical hardware already mentioned.[8] Second, there is the 'recomposition of tasks' lying behind the slogans as job-enrichment or job-rotation. This he sees as the consequence of having removed all the heavy manual jobs at one end of the scale, and much of the skilled fitting and wasteful direct supervision at the other. What remains are bland, interchangeable, semi-skilled tasks. The benefit to managements lies in increased flexibility in the allocation of task-combinations to individuals, who all receive a time-wage rather than any piece-wage. This can contribute to solving the 'balance problems' of earlier forms of mechanisation and also eliminate insufficiently controlled pockets of skilled labour or well-organised unskilled labour. It is instructive to compare the picture of the engineering operative which emerges from this analysis of neo-Fordism with the accounts of the process-operator in those parts of the chemical industry which exhibit continuous-flow processes (Nichols and Beynon, 1977). The same dependency on the machine behind an illusion of skill and judgement is suggested, even though the levels of integration of the two processes are different. Neo-Fordism is discussed further in Chapter 5.

The period since World War Two can therefore be summarised as having witnessed two major changes in production systems. The first which began early on and was to some extent emerging during the 1930s and the war itself, was a change in the focus of mechanisation from transfer to control. The tertiary mechanisation which developed in the post-war period added to the maturity of Fordism and to some extent resolved the technical bottleneck which had been present in the secondary mechanisation stage. But the progress of tertiary mechanisation led to a further bottleneck which was present in both the technology and the labour organisation aspects of the production process, namely the lack of flexibility to cope with small batches and therefore, by implication, with product differentiation. The second change then is the development of technologies and forms of organisation which do possess flexibility. This is the core of what has been labelled neo-Fordism. Most agree, however, that this tendency is still very much in its infancy. Aglietta, without any reference to long wave theories, sees the current recession as linked to the 'crisis' of Fordism, with the ability of neo-Fordism to resolve the situation by no means guaranteed. Mandel (1980, p. 45) although without any technical elaboration, argues that a change in the organisational form of production processes is vital for capital to re-establish sufficient profitability to begin a new long upswing. Finally a wide variety of authors who believe that the current recession has some peculiar 'structural' character to it believe that new technology will be somehow implicated in any upswing. There are several suggestions then, that long periods in capitalist development are related in some way to major changes in the level and character of mechanisation. Further elaboration of this issue must await exposition of long wave theories in the next chapter.

It is appropriate at this point to reconsider another technical development, often associated with 'automation' in the public mind, but not considered either by Bell, Aglietta, Bright, Mandel or any other of the authors we have so far discussed. The industrial robot, perhaps because of its anthropomorphic character, exemplifies the popular concept of 'automation' more powerfully than do numerically controlled machine-tools. The fact that their diffusion and technical maturity have accelerated in the past two or three years leads to the suspicion that they are involved in some way in the developments we have discussed.

The first interesting fact to note about robots is that their practical development was closely linked to that of NC machines. An early example was a tool-changing device in a machine tool, and the whole technical apparatus of electronic control is broadly similar to NC control (Atkinson, 1980, Chap. 5). From the standpoint of Bell's scheme, as used here, a problem arises immediately in that robots are engaged both in transformation operations, such as painting and welding, and in transfer operations. These two types of robot cannot therefore be classified together and have to be discussed separately. In order to retain continuity with the previous discussion of machining it is useful to discuss *transfer robots* first.

Some of these robots, particularly those known as 'pick-and-place' robots, are in fact remarkably unsophisticated by comparison with their popular image. While they do represent level 4 in Bell's hierarchy of transfer, Atkinson argues (1980, p. 174) that they only lie at level 6 in Bright's hierarchy of mechanisation (see Figure 2.1). Perhaps the simplest way to think of this sort of transfer robot is that it is one step up from the rigidity of the handling aspects of a conventional Bell level 4 system as in a transfer line. When operating it is 'dedicated' and 'hard'. It can only work on one or a small number of parts and only make one movement. But it can in principle be re-programmed, and re-tooled (in the sense of being given a new gripping system) so that it can function with different work-pieces. This makes it more flexible than the transfer parts of a transfer line.

However, if we jump to levels 16 and 17 of Bright's chart, levels which were not highly populated by examples in his time, we find robots now in existence with sensory systems and slightly more flexible 'hands'. These robots can cope with the handling tasks of a highly mechanised machining system which has a non-uniform flow of varying parts coming through it. It therefore adds a further level of handling to Bell's scheme.

These robots can now be more clearly situated with respect to the earlier discussion. The progressive mechanisation of control functions currently going on is resolving some problems but is simultaneously exposing the rigidity and incompatibility of existing transfer systems. Complex vision and learning algorithms in the next generation of handling robots will create new degrees of flexibility in handling which will be necessary for the full exploitation of the potential of DNC machining. In fact, such combinations are the basis of US and Japanese experiments in so-called 'workerless' factories. It should be added that these experiments also involve the use of two related technologies: the central process control computer, which controls material and production flow, and the assembling robot, which is a similar device to the level 17 transfer robot but with different grippers and some ancillary jigs and associated hardware.

This leads us naturally on to a brief consideration of *non-transfer robots*. One category is the *assembly robot,* just discussed, the other is the *transforming robot* which does such tasks as painting and welding. What distinguishes these transforming robots is the simultaneous high-level mechanisation of the transformation and control dimensions of tasks which had previously been at the hand-held power tool level. They therefore represent quite large jumps in the degree of mechanisation of these processes. We have already noted that the progress of mechanisation is uneven in various industries and tasks, with not all parts of the production process going through the successive phases of transformation, transfer and control dominance in mechanisation. These examples seem to be laggards which are being catapulted into a higher level of mechanisation by the introduction of a type of technology which exhibits the new 'holy grail' of control mechanisation: flexibility.

It seems in order, therefore, to add robotics, and large-scale computer process control, to the two technological components of neo-Fordism outlined earlier. They form a package by virtue of their shared emphasis on control and on solving the problems encountered in Fordism.

This then raises the question of how far the transition to these new production systems has proceeded up to now, what factors will govern their future diffusion and how long the complete transition it will take. The theoretical answers to these questions depend on the interpretation of the central issue of the relation of technical change to economic development. However, one enabling condition can already be noted as severely under-developed. Although the micro-electronics components which are at the centre of these new technologies are at a reasonable stage of maturity, the sensing, actuating and gripping technologies, as well as the software associated with them, are still in an early stage of development. While technical solutions to these problems may indeed be imminent, we can at least be sure that there are *not* a whole legion of mature industrial robots and computer-aided manufacturing systems which are now entering their *rapid* diffusion phase. It seems more realistic to characterise them as being in the gently-sloping early part of the classic S-shaped diffusion curve, which of course is a difficult portion of the curve to extrapolate.

CONCLUSION

The initial question posed in this and the last chapter was that of identifying 'automation' analytically and establishing its historical character. After noting the great confusion which has surrounded the term in the past, some order was achieved by postulating levels of 'automation'/mechanisation in the manner proposed by Bright. His categories however, tend to blur an important distinction, pointed out by Bell, between the transformation, transfer and control elements of a process. So we have abolished 'automation' as an analytically useful term and chosen to distinguish between different types of mechanisation. We have adopted this scheme, and examined its relationship to the Marxist notion of the labour process. This has resulted in a view of production technology and work organisation as elements of production processes which simultaneously create products and profits and maintains the labour-management relationship in capitalist societies.

When set against the history of the changes and conflicts over the technical and organisational aspects of production systems, this framework has a useful organising power. Although there are any number of inter-industry differences and nuances of periodisation, we have argued that the development of mechanisation technologies has a discernible structure. Once the basic features of machinofacture were established the first phase was the progressive mechanisation of transformation processes, using improved power sources, and improving their integration by means of interchangeable parts. The progress of the mechanisation of transformation gradually shifted the focus and the inducement mechanism towards transfer functions. The linear flow of work-pieces exemplified by Fordism drew on both the problems of congestion and the possibilities of systematisation created by the previous phase. It also entered into a virtuous circle with the expansion of mass-markets for certain consumer items, especially in the areas of transport, domestic work and entertainment.

Subsequently, the focus shifted again to the dimension of control as the incremental improvements in transfer mechanisation yielded diminishing returns. In the post-war boom, control mechanisation developed substantially, but again a bottleneck was encountered in the form of inflexibility. So in the last two decades best-practice technologies have increasingly been differentiated from their forerunners by greater flexibility with respect to scale, batch size, product-type and skill requirements. The technologies used to achieve this flexibility promise to be sufficiently powerful to feed back into the mechanisation of transformation and transfer raising them to higher levels.

Overlaid on this sequence of changes in mechanisation technologies has been a longer process of transformations in work organisation. During the initial progress of secondary mechanisation, Taylorism and systematic management were incorporated into Fordism. This system of production matured and prospered in conjunction with mass consumption, and was fuelled by the successful overcoming of the technical problems of secondary mechanisation and the shift to tertiary mechanisation. But the forms of tertiary mechanisation installed in production processes since World War Two proved insufficiently flexible and became entwined in the problems of Fordism.

New technologies, based on micro-electronics and computers, now beginning their industrial diffusion, offer much greater flexibility in tertiary mechanisation. The significance of this is twofold. One is the extension of mechanisation to small-batch production which previously had to rely on craft workers using multipurpose machinery to make the variety of products associated with batch production. We will show in Chapter 6 that new technologies have major implications for the productivity of small-batch sectors. However, a second effect of flexible tertiary mechanisation is seen within *high volume* production. Here the flexibility of reprogrammable machine systems allows increased mechanisation but with substantial product differentiation. The link between mechanisation and scale begins to weaken; hence, the economic importance of new control technologies in large volume- but not mass-production industries like earth-moving equipment and mid-market furniture. Crucially, computer-based control systems have the potential to cheapen the cost of capital equipment because of reprogrammability. No longer does the machinery on a mass-production line have to be scrapped upon a change of model or product. Its economic life can be extended by reprogramming, thus increasing the opportunity to amortise the cost of this machinery over the life cycle of several production periods.[9]

We therefore now see the onset of a period in which substantial change is imminent in technology and in work organisation. There are also important implications for the type of industrial and, as we shall show in Chapter 7, service products that will be available from these changed production systems. But before we discuss these changes we conclude by noting that the periods of change we have outlined – our phases of mechanisation – bear a striking similarity to the broad structure proposed for long wave interpretations of the history of economic growth first mentioned in Chapter 1. This topic is one which we wish to pursue. In order to do that we now turn to an account and critique of some of these theories.

4 Mechanisation and Long Waves

INTRODUCTION

Chapter 1 drew attention to the recent interest in theories of long-term economic development which might throw some light on the causes of the current economic crisis and on means of solving it. Such theories are usually called 'long wave' theories.[1] There are three reasons why we examine long wave theories in this book. First, they offer, more than any other body of economic theory, a way of examining the historical significance of major technological change. Second, the proposed dating of them corresponds at various points to our analysis of phases of mechanisation. Third, they themselves propose a role for 'automation' which deserves attention, given our focus on mechanisation.

The most developed long-wave theories are either Marxist, such as that of Mandel (1975) or neo-Schumpeterian, such as those of Van Duijn (1983), Mensch (1979) and Freeman *et al.* (1982). It might be prudent to argue, however, that at the present stage of development of such a large-scale theoretical structure it would be unwise to restrict discussion entirely to any one economic paradigm. This is especially true of a situation in which the empirical data which might illuminate long-wave theories are very sparse, and open to many interpretations. Our approach is close to that described by Freeman as 'reasoned history' (Freeman *et al.*, 1982, p. ix). In this chapter we shall critically discuss the work of Mandel and Freeman, concentrating on the way they treat mechanisation in their theories. We have two goals: first, we offer some data on the diffusion of mechanisation technologies since World War Two which 'tests' the model of mechanisation development put forward in Chapter 3 and substantially modifies Mandel's account of them; second, we suggest how our emphasis on mechanisation and labour-process developments contributes to an understanding of the nature of the lower turning point of the long wave.

MANDEL'S THEORY OF LONG WAVES

Mandel's starting-point is the Marxist explanation of the mechanism of industrial cycles in capitalism. He poses the question: 'Is this cyclical movement simply repeated every ten, seven, or even five years? Or is there a peculiar inner dynamic to the industrial cycles over longer periods of time?' (Mandel, 1975, p. 110). His answer to this question is particularly interesting.

The length of the cycles in Marx's analysis is determined by the duration of the turnover time necessary to replace fixed capital, this being something which takes place over a number of successive production periods. In the reinvestment phase a portion of the surplus created is reinvested, thus increasing the stock of capital and constituting what Marx called extended reproduction – the accumulation of capital. In the process of reinvestment, individual capitalists have to make judgements about the sort of machinery in which they should embody their fixed capital. This involves considering the relationships between the technological characteristics of competing machines and other commercial factors which, taken together, determine the costs of production, the levels of output and employment, and the profits which might accrue. In a competitive environment, enterprises will be attracted to techniques in which 'the saving on *paid* living labour exceeds the additional costs of the fixed capital, or more precisely, the total constant capital' (Mandel, 1975, p. 111). We might rephrase this condition as expressing what could be called the need to improve cost efficiency *internally*, that is, as measured with respect to the previous practice of the enterprise. Enterprises will also be attracted to those techniques which not only save labour but also reduce *total* production costs in such a way as to create an extra margin of profit as compared with other enterprises. This second condition expresses the pressure to adopt what conventional economists would call the 'best-practice technique', or to innovate – that is, to bring in a new best-practice technique. On balance, therefore, reinvestment would involve the use of a 'higher' level of technology and an increase in the organic composition of capital which corresponds very roughly to an increase in the capital intensity of production.

Now this model of 'choice of technique', which assumes that enterprises trust their ability to make accurate cost forecasts, implicitly requires a continuous spectrum of techniques to be available – continuous in the sense that for each technique there is a nearby one which just slightly alters the calculations of production-cost advantage versus capital-cost penalty. Thus, the enterprise will be able to optimise (if not maximise) revenues by choosing 'just the right' technique. In reality, of course, technological possibilities are not so well-ordered and Mandel suggests that in general there might be two broad types of reorganisation of production open to the enterprise during reinvestment. On the one hand there are relatively routine 'incremental' changes, where minor improvements are found in the new machinery, labour is more efficiently organised and perhaps some raw materials are economised or cheapened. On the other hand, there may be much more 'radical' changes, involving completely different machinery and work organisation, new raw materials and a 'fundamental revolution in technology' (Mandel, 1975, p. 112).[2]

While it is plausible to construct such a scenario of technical possibilities, the problem remains of how the choice is made in practice and what enabling circumstances are necessary to realise each of the alternatives. Mandel suggests a mechanism for this process with two components. First, he argues that in the course of a normal short industrial cycle the total mass of surplus-value generated over the course of the (approximately) ten-year period exceeds the sum of the surplus consumed by capitalists and reinvested in circulating capital, by some increment, which is available for new fixed capital formation. But the increment associated with *one* short cycle is unlikely to be adequate to create the new production sites for the manufacture of those radically new machines which will be involved in his 'fundamental technological revolution'. Thus, says Mandel, the increments of a succession of short cycles are incompletely invested, and an historical reserve fund of capital is generated which eventually can, 'at the appropriate time', provide the means for the additional accumulation of capital which is entailed when extended reproduction occurs in the context of a 'fundamental renewal of production technology' (Mandel, 1975, p. 114).

The determination of the 'appropriate time' is the second component of the mechanism. If a component of total social capital has been lying 'idle' for a period, this is probably because it could not be invested earlier at an adequate return. Reinvestment then proceeds when an adequate rate of profit looks possible. The accelerated accumulation of the upswing of a long wave therefore depends upon the appearance of factors which create an increase in the social average rate of profit, beyond that which occurs as a

TABLE 4.1 *Long-wave movements according to Mandel*

Long Wave	Main Tonality	Origins of this Movement
1 1793–1825	expansive, rising rate of profit	Artisan-produced machines, agriculture lags behind industry–rising prices for raw materials. Fall in real wages with a slow expansion of the industrial proletariat and mass unemployment. Vigorous expansion of the world market (South America).
2 1826–47	slackening, stagnant rate of profit	Dwindling of profits made from competition with pre-capitalist production in England and Western Europe. Growing value of capital neutralises the higher rate of surplus-value. Expansion of the world market decelerates.
3 1848–73	expansive, rising rate of profit	Transition to machine-made machines lowers the value of capital. Massive expansion of the world market following the growing industrialisation and extension of railway construction in the whole of Europe and North America, as a result of the 1848 Revolution.
4 1874–93	slackening, rate of profit falls, then stagnates, then rises slightly	Machine-made machines are generalised. The commodities produced with them no longer produce a surplus-profit. The increased organic composition of capital leads to a decline in the average rate of profit. In Western Europe real wages rise. The results of the growing export of capital and the fall in the prices of raw materials only gradually permit an increase in capital accumulation. Relative stagnation of the world market.
5 1894–1913	expansive, rate of profit rising, then stagnant	The capital investments in the colonies, the breakthrough of imperialism, the generalisation of monopolies, profiting even further from the notably slow rise in the price of raw materials, and promoted by the second technological revolution with its accompanying steep rise in the productivity of labour and the rate of surplus-value, permit a general increase in the rate of profit, which explains the rapid growth of capital accumulation. Vigorous expansion of the world market (Asia, Africa, Oceania).

6 1914–39	regressive, rate of profit falling sharply	The outbreak of the War, the disruption of world trade, the regression of material production, determine growing difficulties in the valorisation of capital, reinforced by the victory of the Russian Revolution and the narrowing of the world market which it provoked.
7 1940/45–66	expansive, rate of profit first rising, then slowly starting to fall	The weakening (and partial atomisation) of the working class determined by fascism and the Second World War permit a massive rise in the rate of profit, which promotes the accumulation of capital. This is first thrown into armaments production, then into the innovations of the third technological revolution, which significantly cheapens constant capital and thus promotes a long-term rise in the rate of profit. The world market shrinks through autarky, world war and the extension of non-capitalist zones (Eastern Europe, China, North Korea, North Vietnam, Cuba), but is then significantly extended by the intensification of the international division of labour in the imperialist countries and the beginnings of industrialisation in the semi-colonies.
8 1967– . . .	slackening, rate of profit falling	The slow absorption of the 'industrial reserve army' in the imperialist countries acts as a block to a further rise in the rate of surplus-value despite increasing automation. The class struggle attacks the rate of profit. The intensification of international competition and the world currency crisis work in the same direction. Slow-down in the expansion of world trade.

SOURCE *Adapted from Mandel (1975)*.

natural fluctuation in the short cycle. Under these circumstances the under-invested capital of the preceding succession of short cycles can be channelled into the creation of the new branches of production which furnish the capital goods characteristic of the fundamental revolution in productive technology.[3]

Mandel suggests some factors which might enable such a sustained increase in the rate of profit to occur. These are:

1. a sudden fall in the 'average organic composition of capital' which might be seen also as an increase in capital productivity (cf. Soete and Dosi, 1983);
2. a sudden rise in the 'rate of surplus-value' as a result, for example, of a radical defeat of the working class which lowers the rate of wage rises even in a period of economic prosperity;
3. a sudden fall in the price of the elements of constant capital, especially of raw materials, which has the same effect as (1) above;
4. a sudden abbreviation of the turn-over time of circulating capital because of the perfection of new systems of transport and communications, improved methds of distribution, accelerated rotation of stock and so on (Mandel, 1975, p. 115).

Mandel sees a need for several of these 'trigger factors' to be coincident if the increase in the rate of profit is to be sustained, and not damped in a short time-span by a flow of capital into the new areas. This is an interesting part of Mandel's analysis. He insists that the trigger factors are relatively autonomous and not entirely dependent on the pre-conditioning of the downswing of a previous long wave. They are in many senses historically contingent factors which, while related to the economic circumstances of present and preceding periods, are nevertheless related in a very mediated way. He therefore attempts to straddle the ground of a traditional long-wave debate which began in the 1920s, in the USSR, between Trotsky and one of the first long-wave theorists, Kondratiev, over the role of endogenous and exogenous factors, between 'wave' and 'cycle' (see Day, 1981). Mandel sees the response of the capitalist system to major stimuli as law-bound and cyclical but the stimuli themselves are not necessarily regular in occurrence and therefore the cycles appear not as regular cycles but as less regular waves.[4] To underline this historical specificity of the trigger factors, Mandel gives a brief outline of what he believes them to be for the upswings of 1850, 1894 and 1945. These are summarised together with analysis of movements in the rate of profit, components of capital etc. in Table 4.1.

We can therefore characterise Mandel's theory as a modified capital accumulation model. Long waves are seen as alternating periods of accelerated (upswings) and decelerated (downswings) accumulation of capital in which the facts of acceleration and deceleration are the result of the dynamics of a competitive economy given certain boundary conditions. The timing of the turning points, however, is the result of the interplay between the rate of profit and the gestation periods of fundamental innovations, both in technological and money terms. Recently however, Mandel (1980, 1981) has placed more emphasis on the relatively autonomous role of class struggle in the determination of the lower turning point.

Mandel sees the 'revolutions in technology' mentioned above as having some broad structural features in common with each other, in that they are concerned with the technologies of motive power. Thus the three technological revolutions experienced since the middle of the nineteenth century are summarised as:

> Machine production of steam-driven motors since 1848; machine production of electric and combustion motors since the 90s of the nineteenth century; machine production of electronic and nuclear-powered apparatuses since the 40s of the twentieth century. (Mandel, 1975, p. 118)

The upswing of each long wave is characterised by the establishment of the new production sites and new industries, responsible for the production of the new capital goods. This is the activity which absorbs the previously idle capital, giving a stimulus to accumulation. The downswing is characterised by stabilisation of the new sectors coupled with the gradual generalisation of their products, the new capital goods, into many sectors of industry. With the removal of the dynamic impetus of new capital formation in the capital goods sector, the economy as a whole moves to a phase of decelerated accumulation. Clearly, if the new capital goods are relatively labour-displacing, then their application, without the counterbalancing formation of new capital in the capital goods sector, will lead to unemployment in aggregate, without changes in the distribution of wealth, thus compounding a slow-down in economic growth. This aspect of the wave is however dealt with more effectively by Freeman and we will return to it.

Mandel's theory of long-term economic development thus emphasises technical change in the capital goods sector as being a very significant factor. Two other long-wave theories also consider technical change. First, there are the 'innovation-clustering' theories of Mensch (1979), Kleinknecht (1981a, 1981b), and van Duijn (1983). Second, there are the theories which emphasise the clustering of *particular,* and particularly important, technological innovations which form 'New Technology Systems' and generate a collection of new-industry life cycles, or which 'rejuvenate' older, mature industries. Freeman (1977), Freeman *et al.* (1982) and, to a lesser extent, van Duijn fall into this category. We can now consider each of these in turn and examine their interactions with the theory of labour process changes developed in Chapters 2 and 3. One of the features of comparison between the theories which we wish to stress is the relative weight they place on the role of product versus process innovations in their explanation of the ups and downs of the long waves.

THEORIES OF THE CLUSTERING OF TECHNOLOGICAL INNOVATIONS

Mensch takes as his principal focus the clustering of *basic* innovations and tries to establish it as a documented and statistically verified phenomenon. Thus a large part of his work is devoted to tables and charts with his estimates of the dates of inventions and innovations. From this Mensch invokes elements of Schumpeter's economic analysis and explains the great depressions of economic history as points of 'technological stalemates' when the industries created by basic innovations have run out of steam, no longer stimulating the economy as a whole. The process of running out of steam has its technological counterpart in the stream of incremental improvement innovations which follow radical innovations. The stream of improvement innovations coincides with the evolution of the relevant industry along the path of the product's life-cycle.[5]

Mensch does not believe that this process is the result of any discontinuity in the availability of basic inventions to be innovated, although he criticises business practice for neglecting investment in research and development in such a way as to reduce inventive as well as innovative possibilities. He uses his statistical samples and techniques to argue that the basic inventions are much more equally distributed in time than are the innovations (Mensch, 1979, p. 148). There is some gentle clustering but only as a reflection of very long-lived changes in basic scientific theory. He sees the clustering of innovation as resulting from a compression of the invention–innovation lead-times as a result of the pressures of the recession.

This view of industrial development – inventions continuously generated, innovations clustering – bears a close resemblance to that of Schumpeter. The burden of economic growth-stimulating innovation falls upon the activities and perceptions of individuals and particularly upon individual entrepreneurs. Around the upper turning point, the product differentiation, planned obsolescence and gimmicky competition of mature consumer-goods industries play a part in the stagnation of demand by producing dissatisfaction amongst consumers. During this period, Mensch argues, consumers are reaching towards new demands which the supply side cannot satisfy thus compounding depression. Delbeke (1981) has correctly pointed out that this emphasis on demand saturation is a feature not given such prominence in other long-wave theories.

Thus, for Mensch, the downswing of a long wave is best seen as an innovation 'bottleneck'. In the last resort, entrepreneurs adopt new basic innovations because it is the only alternative left, save complete commercial suicide through pursuing a product line which is completely stagnant. We shall see later that this view of the effects of recession on technical change is dubious theoretically and not yet empirically proven. Mensch's work is therefore more useful in its emphasis on the role of product technology in upswings than in its analysis of turning-points. Further work on clustering by Mensch and other authors (Kleinknecht, 1981a, 1981b; van Duijn, 1983) has been useful in clarifying the debate, but we are more convinced by Freeman's treatment of this problem and it is that to which we now turn.

'NEW TECHNOLOGY SYSTEMS' THEORIES

The work of Freeman and his colleagues is written in the context of the debates over structural unemployment, and attempts to explain that phenomenon as intrinsic to the downswing of a long wave, arguing that this is the case in the last part of the twentieth century. Freeman's starting point is Kuznets' request for a clear statement of the relationship between the rate of introduction of a major innovation and the upswing of the long wave. He argues that what is necessary is to investigate the possibility that:

TABLE 4.2 *New technology systems: a schematic representation*

	Developments in depression phase of previous wave	Developments in new wave		
		Recovery and boom	Stagflation	Depression
Research, invention	Basic inventions and basic science coupled to technical exploitation. Key patents, many prototypes. Early basic innovations.	Intensive applied R & D for new products and applications, and for back-up to trouble-shooting from production experience. Families of related basic innovations.	Continuing high levels of research and inventive activity with emphasis shifting to cost-saving. Basic process as well as improvement inventions are sought.	R & D-investment becomes less attractive. Despite the fact that firms try to maintain their level of research it becomes increasingly difficult to do so with the slackening of their sales. At the same time the volume of sales required to amortise the cost of R & D is steadily increasing. Basic process innovations still attractive to management but may meet with social resistance.
Design	Imaginative leaps. Rapid changes. No standardisation, competing design philosophies. Some disasters.	Still big new developments but increasing role of standardisation and regulation.	Technical change still rapid but increasing emphasis on cost and standard components.	Routine 'model' type changes and minor improvements of cumulative importance.
Production	One-off experimental and moving to small batch. Close link with R&D and design. Negligible scale economies.	Move to larger batches and where applicable – flow processes and mass production. Economies of scale begin to be important.	Major economies of scale affecting labour and capital but especially labour. Larger firms.	The slowdown in output and productivity growth leads to over-production and excess capacity in some of the modern industries. These structural problems are 'cumulative and self-reinforcing', with repercussions for the economy at large, and lead to a further decline of economic activity.

entrepreneurs. Some large firms. Fairly labour-intensive. Problems of venture capital.	new capacity. Band-wagon effects. Large and small firms attracted by high profits and new opportunities.	shifting to rationalisation. Continuing rapid rapid growth but increasingly large sums required to finance R & D and rising capital costs. Rising capital intensity.	the capital stock in some of the most modern sectors of the economy; low profit margins and the general 'pessimistic mood' with regard to expectations lead entrepreneurs to be very (over-) cautious in relation to new investment opportunities. Investment which will take place will be primarily directed towards rationalisation. Search for new investment opportunities abroad.
Market structure and demand	Innovator monopolies. Strong consumer resistance and ignorance. Some new small firms to promote basic innovations.	Intense technological competition for better design and performance. Falling prices. Big fashion effects. Many new entrants in early build-up.	Growing concentration. Intense technological competition and some price competition. Strong pressure to export and exploit scale economies.	Even stronger trend to oligopoly or monopoly structure. Bankruptcies and mergers.
Labour	Small-scale employment-generating effects. High proportion of skilled labour, engineers and technicians. Training and learning on the job and in R & D.	Major employment-generating effects as production expands. New training and education facilities set up and expand rapidly. New skills in short supply. Rapid increase in pay.	Employment growth slows down and as capital intensity rises, some jobs become increasingly routine.	Employment growth comes to a halt. Unemployment rising. In addition to the continuing labour displacement effects of rationalisation investments, employment suffers (in the first instance) from the general recessional and depressional tendencies in the economy at large.
Employment effects on other industries and services.	Negligible, but imaginative engineers, managers and inventors are thinking about them and planning and investing accordingly.	Substantial secondary effects, mainly employment-generating but gradually swinging to displacement.	Labour displacement effects, as new technology now firmly established and strongly cost-reducing.	Continuing labour displacement as new technology penetrates remaining industries and services.

SOURCE *Adapted from Freeman et al. (1982).*

(the) early phase of the introduction of major new technologies might tip the balance of economic development in an expansionary 'full employment' direction, whilst later phases of the exploitation of these same new technologies might tend to shift the balance in the opposite direction (Freeman, 1977, p. 7).

Whilst acknowledging the stimulus of Mensch's work on the cataloguing of innovations, Freeman prefers to concentrate on a very few fundamental changes in technology which embrace a large cluster of associated or dependent specific innovations, called New Technology Systems (NTS). Freeman feels that in the post-World War Two period electronics and synthetic materials can be seen as such systems. Whilst he does not wish to propose that such major technological changes be regarded as the single cause of the long waves, he clearly feels that at least some of them would be likely to be synchronised into the waves by virtue of their structural interactions with the economic system. In order to have the profound effects on the economy which the theory suggests, they are most likely to be associated with the machinery industries, transport, energy and communications.

A model is developed which attempts to show how the sequence of events in research, development, design, application, marketing and employment effects of an NTS might be logically 'fitted' into the long-wave schema. A version of this schema is shown in Table 4.2 and can be summarised as follows:

1. Major new technologies begin to appear throughout the wave preceding the one in which they boom.
2. In the upswing, the new technologies establish new branches of industry which exert a demand–pull stimulation on supplying industries and which contribute 'spin-off' applications which are taken up by these industries. Relative prices of the new goods fall rapidly.
3. The effect of (2) on employment is to create new jobs in new and closely connected industries. Skill shortages and full employment allow wage-rates to rise, creating inflation and long-term pressure to adopt more labour-saving techniques.
4. This employment-generating effect dwindles as the growth rates of the new branches subside and the focus on rationalisation develops. This stimulus to the whole economy therefore subsides also.

5. Any labour-displacing potential of the new technologies is seized upon and utilised fully in the downswing, thus exacerbating demand-deficiency type constraints on economic growth.[6]

In some respects this model is not inconsistent with Mandel's if we abstract from the Marxist conceptual structure of value categories. But there is a major difference in that Mandel sees 'automation goods' as the major innovations of the current wave, while Freeman emphasises a general trend toward labour-saving in all downswings and sees other innovations as the impetus to upswing. A second difference is Freeman's emphasis on the labour market as a source of 'stickiness' helping to generate the disequilibrium of the long wave. We will now examine more closely the difference between the arguments of Freeman and the clustering theory of Mensch.

Freeman has criticised the data, their interpretation, and the theoretical basis upon which Mensch bases the notion of a 'depression trigger' to innovation. He does, however, strongly believe that basic and fundamental innovations exhibit clustering, and with his colleagues has substantiated this with patent and invention data (Clark *et al.*, 1981). But in their view the clusters are not necessarily located in the depression phases of long waves, as Mensch believes. Nor do they find any evidence of increased attempts at innovation in a depression; rather they argue that R & D in these circumstances tends to be curtailed.

Freeman's explanation of the long wave turns not on the clustering of the innovations but on the clustering of the *diffusion processes* of the technically related products and processes of the new technology system. In fact, the timing of the actual innovations is pushed to the fringes of the picture in a classically Schumpeterian manner. They are not entirely exogenous, but they are certainly more contingent and not linked in a formal manner to the long wave. The approach is to create a model of the possibility of new sectors being created by a family of innovations entering their diffusion curves in a related way. The evolution of such a phenomenon is then traced using historical evidence to illustrate and substantiate the model. Finally, the macro-economic upswing is deduced as a potential corollary as is its upper turning-point and the character of the recessionary period. This leaves one big gap. The moment and nature of the lower turning-point is not included in the model. Freeman clearly feels that socio-political factors are important at this point but he does not elaborate on this.

So what is the general sequence of events over the course of a long wave? First, a number of basic innovations enter the beginning of their diffusion. Eventually this creates the need for new manufacturing processes and so demand and innovative activity in relation to the capital goods sector are increased. Productivity growth, market growth and, as with Schmookler (1966), demand-led innovation now become prominent features of the sector in the rapid growth phase. Eventually, however, labour shortages, cost pressures and increasing competition generate a bias in process change towards labour-saving. With the onset of recession, the absence of new product innovation and the further increased pressure on costs pushes the bias further toward labour-saving technical change. R & D is likely to concentrate on such innovations during this period.

This scheme suggests that the pressures to alter the technical and organisational aspects of the production process will be of two types. In the upswing, according to Freeman, new production processes will be created in new (and in existing) sectors in conditions where potential labour resistance is either bought out, turning into inflationary pressure, or else is alleviated by other aspects of macro-economic growth. In the downswing the picture is quite different. Despite the existence of a labour surplus, technical and organisational change will be intensified in order to reduce costs further and reshape both jobs and processes to meet the perceived needs of managements in search of productivity and quality improvements. This aspect of Freeman's account has features in common with the picture of successive periods of intensified labour process conflict which have been described in Chapter 3.

The broader socio-political nature of the factors affecting change in the downswing is acknowledged by Freeman. He points out that while disagreeing with Mensch's notion of the depression trigger phenomenon, he believes that the extremes of social distress and conflict possible in the downswing could cause an 'indirect' depression-trigger phenomenon. In the light of our analysis we can elaborate this view. The barriers to radical change in production can be lowered by the desire to break the log-jam in the economy at a social or political level. Clearly, there are some forms of labour process change which are facilitated by political change as was shown starkly in the German economy under Nazi rule, and in different ways in the wartime economies of the allied countries.

Indeed the role of World War Two in the lower turning-point of the most recent long wave is clearly so dominant that Freeman, with good reason, finds it difficult to suggest any more mechanistic or purely economic structure to the turning-point. The influence of rearmament was important from the mid-1930s onwards. In the presence of such an overwhelmingly socio-political and contingent feature as a world war it is a brave theorist who insists on primacy for some underlying economic movement in explaining two decades of movement out of depression into prosperity. Nevertheless, the war was clearly connected to the broader circumstances which Freeman cites as possible components of an 'indirect depression-trigger' effect on technical change. It seems uncontroversial to point out that there was a connection, if complex, beween the political trends which were important in the eventual collapse into war and the class conflicts resulting from the economic circumstances of the 1902s and 1930s. That these conflicts had a component which centred on the manner of the functioning of production processes also seems plausible.

The basic issue at stake is the rigidity or contingency of the economic fluctuations. At one extreme is the view that there is some natural necessity to the long-wave phenomenon. At the other extreme it is possible to formulate a picture in which the socio-political dimensions of the depression so confuse and override any pure economic or technical relationships that we are placed firmly in the field of contingent history where outcomes of complex situations are not predetermined. This dispute is the same one which has always been at the centre of the discussion of long waves: do the long waves behave as *cycles,* or as successive periods which have sufficiently different characteristics to make it impossible to regard them as cycles? We have made our position on this point clear earlier. The long wave is relatively 'law-bound' once an upswing is under way, but the factors which precipitate an upswing are not solely economic.

We cannot resolve this dispute here. What we have emphasised is the likelihood of a connection between production-process change and long waves. The focus of this connection is at the lower turning-point, which is the most incomplete part of the long-wave theory. All recent writers are agreed that political and institutional changes are the key to the initiation of an upswing with techological change as a further condition which is necessary but not sufficient. Gordon, Edwards and Reich (1982) for example, have attempted to document the institutional change at lower turning-points of long waves in the USA (and we return to their arguments in Chapter 8). Perez has argued that:

The structural crisis thus brought about is then, not only a process of 'creative destruction' or 'abnormal liquidation' in the economic sphere, but also in the social-institutional. In fact, the crisis forces the restructuring of the socio-institutional framework with innovations that are complementary to the newly-attained technological style or best practice frontier. The final turn the structure will take, from the wide range of the possible, and the time span within which the transformation is effected to permit a new expansionary phase will, however, ultimately depend upon the interests, actions, lucidity and relative strength of the social forces at play. (Perez, 1983, p. 8)

Freeman (1983b) has shown that Keynes, despite having no particular view of long waves, agreed with Schumpeter that major surges of investment would be fundamentally conditioned by the combination of technical opportunities and investors' profit expectations. All these writers then, tend to converge on the expectations of the rate of profit, and its political determinants, as the barometer of the lower turning-point, thus adding further substance to the argument of Mandel on this point. We suggest that transitions between phases of mechanisation have also been intertwined with the volatile combination of events in the downswing and lower turning-points. The conflict over how production processes should be organised, if resolved, can make a significant contribution to the formation of profit expectations. The resolution may be primarily technological or organisational or, more likely, some combination of the two. In the last three lower turning points the combinations have been different, but mechanisation regimes have been significant contributors.

Thus we conclude that while the role of technological change in products may be adequately explained in Freeman's framework, the role attributed to it in process change is too passive. In Mandel's framework product change is ignored and process change is inaccurately specified. The problem comes down to his understanding of 'automation'. In the rest of this chapter we seek to put this right.

'AUTOMATION' AND THE POST-WAR BOOM: SOME DATA

It was noted earlier that Mandel's account of technological revolutions in production places 'automation' at the centrepiece of that revolution associated with the post-1945 upswing. There are several problems with this view. Perhaps the most important is that Mandel has no real explanation of why that particular collection of innovations should take place at that time. While he gives some account of pressure on the demand side, in the form of intensified competition over unit costs, he has nothing to say on the supply side. He makes no comment on what specifically where the obstacles to increased productivity with previous techniques, or whence the new techniques came. It is this problem which we wish to resolve.

Mandel defines 'automation', following Rezler (1969), as the application of continuous-flow production methods, the automatic transfer of parts between successive production processes, and computer or electronic control of processes. Needless to say, these three principles are often found combined in practice. Mandel gives only anecdotal evidence as to the increasing use of these techniques in the 1950s and 1960s. This definition of 'automation' is marked by the inadequacies of all the other definitions which were proposed in the debates of the 1960s. It attempts to combine the transfer and the control dimensions of mechanisation and is confused as a result. While it is accurate to say that continuous flow, mechanical handling and electronic control were diffusing in the post-war period, they were not doing so at the same rate nor for the same reasons.

This can be demonstrated by an analysis of Census of Production data collected by the governments of both the USA and Britain at various intervals. These unique and vastly complex surveys are the only direct, disaggregated source of data on what is actually *made* in the economy. In particular, the Censuses list the values of output for each category of product for each Census year. Each category of product in the 'machinery' sectors can be examined in turn to make some judgement on whether or not it might constitute an element of the process of increasing 'automation' alleged to have been taking place during this period. How can this decision be made? We have used two interrelated sets of criteria.

The first set of criteria is that of Bell's axes of transformation, transfer and control. In Chapter 3, 'automation' as a general historic phenomenon was presented as a progression in which each of these dimensions of mechanisation had been successively brought into play in leading sectors. In different industries however this progression has followed a different pattern and moves at different rates. On examining a wide variety of machinery, at least some of which is tailored to specific customer industries, we may expect to find revealed different patterns of the balance of transformation, transfer and control mechanisation.

The second set of criteria is more specific to the post-war period. Earlier in this chapter it was established that Mandel's use of mechanical handling, continuous-flow principles and electronic control, despite not being based on any theory of 'automation' accurately captured the fact that, in this period, 'automation' consisted of *mature* transfer mechanisation being reinforced by *nascent* control mechanisation. Since these categories are couched in a more directly technological language, it should be possible and appropriate to search for explicit examples of these types of hardware in sorting through the Census categories.

These two sets of criteria can therefore be used to structure the production of post-war machinery industries into a pattern which helps to evaluate the general model of mechanisation put forward in Chapter 3 and how it links into long-wave theories. (The potential pitfalls in this procedure and inadequacies in the data themselves are discussed elsewhere.)[7]

The task is, therefore, to sort the various types of machinery in the US and UK Production Censuses – and there are more than sixty major types, each with multiple subdivisions – into one of five categories:

- mechanical handling equipment (for example, conveyor belts)
- continuous-flow process equipment (for example, chemical plant)
- control technologies (for example, process-control electronics)
- hard/dedicated 'automation' technologies (for example, cigarette-making machinery)
- none of the above four

The totals in terms of value of machinery produced in the Census years, for each of the first four categories, are then expressed as a proportion of the total value of machinery in that year. The results of this analysis are presented in Tables 4.3 and 4.4 and in Figures 4.1 and 4.2.

One can see from Table 4.3 that, for the USA, mechanical handling increases from 3.6 to 6.1 per cent of the total machinery produced between 1947 and 1972. Hard 'automation' increases from 0.6 to 2.4 per cent. Continuous-flow machinery remains, proportionately, fairly constant going from 6.6 to 6.9 per cent. Control technologies, however, show a large rise from 3.4 to 16.2 per cent.

TABLE 4.3   *Machinery production in USA 1947–72*

	1947	1954	1958	1963	1967	1972
1 Total machinery production ($million)	10019.7	14784.9	20339.2	26969.1	42677.1	53645.5
2 Mechanical handling value as % of total	356.8 3.6	418.6 2.8	1176.0 5.8	1590.5 5.9	2519.9 5.9	3289.4 6.1
3 Continuous flow value as % of total	659.5 6.6	1046.7 7.1	1414.7 6.9	1954.9 7.2	3132.3 7.3	3703.0 6.9
4 Hard automation value % of total	58.3 0.6	114.8 0.8	120.0 0.6	458.1 1.7	940.9 2.2	1293.7 2.4
5 Control value as % of total	340.2 3.4	617.8 4.2	856.1 4.2	3751.1 13.9	6301.1 14.8	8708.3 16.2
2+3 as % of 1	10.2	9.9	12.7	13.1	13.2	13.0
4+5 as % of 1	4.0	5.0	4.8	15.6	18.0	18.6
2+3+4+5 as % of 1	14.2	14.9	17.5	28.7	31.2	31.6

SOURCE *calculated from US and UK Censuses of Production; these tables were compiled from detailed tables presented in Coombs (1982) and Coombs (1984).*

TABLE 4.4   *Machinery production in UK 1954–79*

	1954	1958	1963	1968	1972	1975	1979
1 Total machinery production (£million)	1401.9	1894.0	2214.5	3247.6	4242.6	7686.7	12977.6
2 Mechanical handling value % of total	41.0 2.9	62.9 3.3	113.7 5.1	187.0 5.8	291.5 6.9	619.9 8.1	1088.9 8.4
3 Continuous flow value % of total	147.2 10.5	190.5 10.1	188.3 8.5	247.5 7.6	332.9 7.8	580.7 7.6	903.4 7.0
4 Hard automation value % of total	25.5 1.8	36.7 1.9	55.2 2.5	87.3 2.7	108.5 2.6	202.5 2.6	367.3 2.8
5 Control value % of total	9.5 0.7	30.9 1.6	91.6 4.1	234.9 7.2	593.7 14.0	868.6 11.4	1829.4 14.1
2+3 as % of 1	13.4	13.4	13.6	13.4	14.7	15.7	15.4
4+5 as % of 1	2.5	3.5	6.6	9.9	16.6	14.0	16.9
2+3+4+5 as % of 1	15.9	16.9	20.2	23.3	31.3	29.7	32.3

SOURCE *As for Table 4.3.*

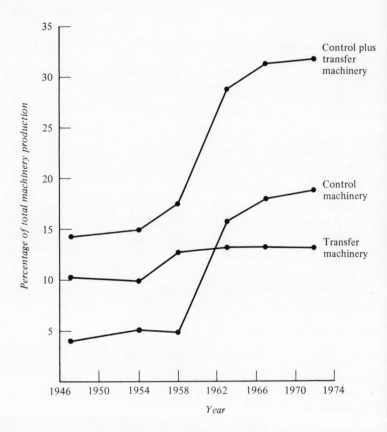

FIGURE 4.1   *Control and transfer machinery production in USA, 1947–72*

This rise is particularly concentrated between 1958 and 1963. For Britain the picture is quite similar to that of the USA. From 1954 to 1979 hard 'automation' shows a gradual increase from 1.8 to 2.8 per cent. Mechanical handling shows a more substantial increase from 2.9 to 8.4 per cent; continuous-flow technologies which were stable in the USA, show a small decline in Britain from 10.5 to 7 per cent whereas control technologies show a very marked rise from 0.7 to 14.1 per cent.

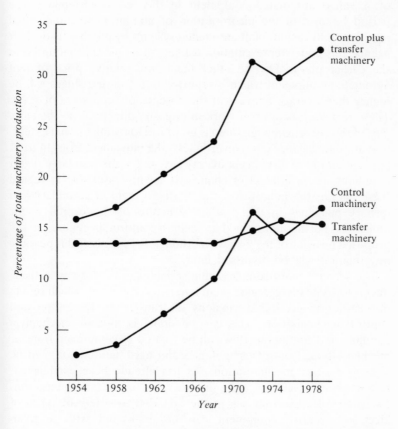

FIGURE 4.2 *Control and transfer machinery production in UK, 1954–79*

Summing all the 'automation' categories as a fraction of all machinery produced gives us a gross indicator of the answer to the dubiously significant question 'how much automation has there been since the war?'. These data are presented in Figures 4.1 and 4.2 and it can be seen that the curves for both the USA and Britain have a characteristic S-shape and rise from around 14 per cent of all machinery in 1947 to around 31 per cent in 1972 for the USA and 16 per cent in 1954 to 32 per cent in 1979 for Britain. It is important here to point out some of the factors which qualify these data and their interpretation. There is likely to be systematic under-representation of so-called automated equipment in the census throughout the period because of the incorporation of non-automated individual machines into custom-built automated systems by purchasers or their agents; this under-representation is likely to be more pronounced in the earlier phase. On the other hand, the relative price of that automated equipment which is reported in the earlier stages will be higher than average because of the absence of economies of scale. These two sources of error work in opposite directions on the early part of the curve; although there is no way of knowing to what extent, if at all, they balance out. Nevertheless, the movement from 14 to 31 per cent and 16 to 32 per cent over twenty-five years is so large that it is unlikely to be artefact or chance. It seems reasonable to regard these data as the indicator of the relative weight of secondary and tertiary machines in the product structure of the capital goods industries. Furthermore, the data seem to conform, in general, to the proposed upswing of the long wave, though the details of this possible interrelation will be discussed later.

The second, more complex, thing which can be done with these data is to try to disaggregate them into two components which reflect our earlier theoretical discussions of transfer mechanisation and control mechanisation. This can be done as follows: mechanical handling and continuous flow can be treated as comprising transfer mechanisation. Control technologies and hard 'automation' can be seen as tertiary mechanisation. As has already been established, tertiary mechanisation is quite distinct from transfer mechanisation. Its empirical distinction lies in the fact that a variety of types of electronic control equipment can be associated with a given production system. While it is true that in real technical change the new equipment will often incorporate an improvement in transfer and control mechanisation simultaneously, it is also true that the control feature can be modified autonomously. It is this relative independence in both conceptual and technological terms which makes control goods valid as a separate indicator from transfer mechanisation goods. Accordingly, in Figures 4.1 and 4.2, transfer and control are shown as two separate components of the time trend of total 'automation' goods. It is clear that the trends are quite different from one another. The transfer mechanisation trend is quite gentle from 10.2 per cent in 1947 to 13.0 per cent in 1972 for the USA, and from 13.4 per cent in 1954 to 15.4 per cent in 1979 for Britain; whereas the tertiary trend is quite sharply rising from 4.0 to 18.6 per cent for the USA and 2.5 to 16.9 per cent for Britain.

Some general observations can now be made on both the US and UK data. First, they are remarkably similar in overall character, both in terms of gross aspect and detail. This is certainly to be expected to some extent since we have attempted to be as near consistent as possible in the choice of categories to assemble the various subtotals. This suggests that it is reasonable to assume that the historical development of mechanisation has been broadly similar in the output of the machinery industries of the two countries. There are also some significant differences however. The UK transfer mechanisation trend is higher than that of the USA over the overlap period from 1954 to 1972. It is rather doubtful whether any significance can be attached to this difference since it is quite small and is broadly accounted for by slight differences in the coverage of the category in the two sets of data brought about by the dissimilar Industrial Classification schemes. In fact, it is likely that any systematic errors in the data will influence equally the absolute value of the percentages in all years rather than the rate of change with respect to time. It is, perhaps, more interesting when looking at the two 'control' curves to note the slope of the steep-rise portions, rather than the fact that the US one is higher than the UK one. Since these technologies are the ones that are diffusing rapidly over the study period, it is important to note that diffusion seems to have accelerated earlier in the USA than Britain. This judgement must be qualified by the fact that the Census category changes, referred to earlier, may have occurred at different rates in the two countries. On inspection it transpires that a separate Census category for computers – a dominant element of the control subtotal – was introduced in the UK in 1958, whilst in the USA the category appears in 1963. This would bias the data in the opposite direction to that observed and suggest that the observed diffusion lead of the USA in all control technologies is real and not an artefact.

There is also a significant difference between the two countries' Control and Total curves in the latter part of the time period. The US curves show evidence of retardation of growth of secondary and tertiary mechanisations' share of machinery output from 1963 but we have no data beyond 1972. In the British case, the retardation comes in 1972 but recovers by 1979. The significance of this is not clear.

Nevertheless the data seem to support the hypothesis advanced earlier; the immediate progress of 'automation' since World War Two has not been one undifferentiated process, but rather the sum of two distinct trends. The first trend is the 'mature' phase of the diffusion of transfer mechanisation and the second trend is the early stage of diffusion of control mechanisation.

CONCLUSION

We have seen that, for the post-war period at least, the data give support to our analysis of phases of mechanisation. Our intention is to incorporate this dimension of technical change into the long-wave model which results from combining Freeman's views on New Technology Systems and product change with Mandel's views on the role of political factors in influencing profit expectations. Mandel's analysis does briefly mention changes in the production process which might occur in the long wave. As part of his emphasis on the relative autonomy of class struggle and its role in the unpredictability of the resolution of the depression, he observes that the depressions are periods of class conflict over the 'method of organisation of labour':

> What we want to stress is not so much the consequences as the *origins* of revolutionary transformations of the labour process. In our opinion they originate from attempts by capital to break down growing obstacles to a further increase in the rate of surplus value during the preceding period. Then, again, the direct connection is established with the rhythmic long-term movement of capital accumulation and the increasing (or decreasing) push toward *radical* changes in labour organisation. (Mandel, 1980, p. 43)

He further argues that Taylorism was a response to the control over the labour process enjoyed by craft workers, and he sees the 'third technological revolution' (automation) as a response to the ability of workers to impose some control on the conveyor-belt production systems which became widespread from the inter-war period onwards.

In the light of the data and the arguments above it is clear that Mandel has been too casual in assuming that his 'technological revolutions' and his brief judgements on the history of the labour process mesh together quite as readily as he suggests. In the post-war period the technologies which Mandel cites as 'automation' and part of the 'third technological revolution' are not, as he suggests, one coherent trend. Nor do they owe their origin entirely to attempts to erode workers' control in Taylorised post-craft processes. In fact, this cluster of 'automation' technologies contains two components. One component is the mature diffusion phase of secondary mechanisation technologies of the *transfer* type. The other component is the initial, rapid growth rate of the diffusion of tertiary mechanisation technologies concerned with *control* in the narrow technical sense. This overlapping of two sets of diffusion processes is the real substance of the capital goods industry expansion which characterises the upswing.

We therefore propose a new component to add to the views of Mandel and the view of Freeman on the possibility of an 'indirect trigger' mechanism in the depression. Our analysis suggests that the depression facilitates a shift from an old 'production regime' to a new one. However, we propose only a shift in the relative inducements and costs of radical innovations in production processes rather than a *necessary* flow of such innovations. It is important to recognise the unpredictability and serendipity in the flow of particular technical solutions to the perceived production problems which are highlighted in a depression.[8] It also follows that the capital goods sector of the economy is a significant economic indicator at the lower turning point of the long wave. Evidence on profit rates and value added in the capital goods industries has shown that an upturn in business expectations in this sector may well be a leading indicator of the long-wave upturn (Coombs, 1981).

Although the conditions which stimulate a shift in production methods intensify in the depression, the diffusion of new techniques, *and* some of the related secondary innovations, are concentrated in the subsequent upswing. The data given illustrate this point. The diffusion of new process innovations was rapid in the post-war upswing. Further, it shows that some innovations, notably those concerned with tertiary mechanisation, were beginning their diffusion almost from zero in the late 1950s and early 1960s, suggesting that some of these innovations themselves were post-war.

The core of our argument then is that a key factor in any explanation of what happens at the lower turning-point of the long wave is what production-process changes are underway, such changes being entwined with the broader political and institutional changes to which long-wave theorists have referred. The production-process changes will be of a technological and work-organisational kind, the technological ones being framed by the pattern of mechanisation developments through the various phases we have described. All the long-wave theories except for that of Mandel tend to relegate process innovation to the role of a secondary, derived feature. Essentially they regard process innovations as contingent on the dynamics of new industry or product life cycles. Freeman does include the possibility that the macro-economic influence of inflationary pressure via the labour market may induce labour-saving technical change but does not make this a major feature of an analysis of the whole wave, only of the upswing and the upper turning-point. Mandel does make process innovation a more central feature of his theory, but no theory of changing *demand* for process innovations is presented; this is only added in the later adjustments and then in a rather incomplete way. For Mandel, technological revolutions are still essentially secondary features. They act as carriers for economic forces unleashed under the stimulus of long-term shifts in the rate of profit.

By examining the features of the development of production processes and their impact on the potential demand for radical process innovations, this chapter has suggested a more subtle, and perhaps more important, role for such process innovations. We need not be forced into choosing to regard process innovations and their diffusion periods as *either* secondary aspects of a prosperity period *or* as serendipitous channels of expansion for the capital goods industry. Instead, changes in production processes as expressed through technological innovation in the capital goods industry have an intrinsic dynamic which is linked into the long-wave mechanism. This link is seen as the micro-economic level in the development of new product sectors and at the macro-economic level through the medium of national labour markets and class struggles.

It should be noted that there is no *necessary* connection in the view outlined here between the technological principles of production process innovations present in a wave and the technological principles present in the *product* innovations which may be important growth centres of an upswing. It is however interesting to observe empirically that there does seem to be *some* degree of overlap between the technologies. Take for example the dual role of electronics in products and processes in the current long wave. This point is one which tends to reinforce the insistence of Freeman on the importance of new technological systems, as distinct from particular individual innovations in the generation of the upswing.

The contribution of the argument developed here is our defence of an *induced* character of much technical change in the production process in the depression and a knock-on effect of this into the upswing. This contrasts with Freeman's view, which tends to restrict demand influences on innovation to explain the upswing, and insists that Schumpeter-like processes dominate innovation in the depression. It contrasts also with Mandel who, until recently, has not seen the origins of the technological revolutions as problematic and has therefore been crudely Schumpeterian on this point by default. Both authors for different reasons struggle to emphasise serendipity (either of science or of politics) at the lower turning-point. Our analysis does not disagree in principle but in degree. The process technical changes of the depression are *not* entirely random in origin or content, but are part of an intelligible sequence of phases of mechanisation of which control has been the most recent. We are not however suggesting that these phases are in any way prime movers for the long-wave mechanism. Nor are we able to suggest *how* these changes do or do not synchronise with the product technical changes which other long-wave theorists choose to emphasise.

The chronology of the phases of mechanisation is summarised in Table 4.5. The last entry in that table reminds us of the discussion of Fordism in Chapter 3 where we argued that the inflexibilities of its production technologies and forms of organisation are now becoming a serious bottleneck. We can now see that this bottleneck coincides with the downswing of the current long wave. This chapter then suggests not only a general result concerning long waves and mechanisation but also a specific one. The proposed emergence of neo-Fordism is a potential component of the lower turning-point of the current long wave. Future chapters are therefore devoted to analysing neo-Fordism in more detail.

TABLE 4.5 *Phases of mechanisation*

	Primary mechanisation	*Secondary mechanisation*	*Tertiary mechanisation*
1850			
	beginning		
1875			
1900	spreading across sectors and maturing technically	beginning	
1925		substantial diffusion in some sectors, increasing technical maturity	
1950	continuing but increasingly likely to occur together with secondary or tertiary	being generalised across a wide variety of industries	beginning in some industries and slowly becoming more flexible
1975			flexibility increasing

5 New Technology and Neo-Fordism

INTRODUCTION

In Chapter 3 we considered the central role of Fordism in the post-war upswing. Fordism, characterised by a technological paradigm of the high-volume production of standardised products, gained its efficiency by means of an acute division of labour along with the use of specialised machinery 'dedicated' to the production of long runs. But as we and others have argued, with increasing competition between firms, the effects of the generalisation of collective bargaining on wage costs and limitations to the exploitation of economies of scale, the potential for further extension and deepening of Fordism, at least within developed countries, has progressively diminished.

Continual spates of mergers and 'rationalisation' of plant and product ranges have been made in an attempt to try to regain the advantages of economies of scale. The sub-contracting of sub-components and modularisation of product assemblies has allowed the backward diffusion of mass-production techniques to suppliers of intermediate goods. As labour-process analysts have demonstrated, an enormous amount of effort has been expended on attempts to intensify the use of labour to counteract falling profitability. Directly and indirectly, social scientists and industrial relations managers have worked on the problem of 'motivations' and 'attitudes', with increased performance and productivity as the criteria for their success.

The limitations of Fordism in providing further productivity increases are even more pronounced in areas of work where volume and standardisation of products and components have proved difficult to attain. Variety in products, processes and raw materials, in addition to the size requirement of markets for products, provides strong barriers to the diffusion of Fordist techniques. Where standardisation of production and consumption cannot easily be ensured, as in the case of small-batch production and some services, productivity growth has tended to be low and such products relatively expensive. In manufacturing, the flexibility of operations characteristic of work processes which exhibit such variety has mainly been ensured by the use of skilled labour. Such flexibility has been an important factor in the continued survival of craft workers. In this and the next two chapters we shall look at the means whereby the barriers to the wider extension of mass-production techniques could be overcome. Writers and management specialists dealing with the problems of Fordism have concerned themselves either with work organisation or with the issue of 'new technology'. We will try to integrate these two areas by analysing the phenomenon of *neo-Fordism* as an emergent paradigm of labour process 'design' involving new forms of control mechanisation. The intention will be to provide a basis for assessing the strategic importance of work reorganisation and technological change in the current restructuring process.

WORK ORGANISATION

We have described how a 'labour process analysis' of work, such as that of Braverman, defines the general trend of work organisation as towards increased division of labour, routinisation and simplification of most work tasks, whilst creating a hyper-skilled technical stratum, whose major function is to design and organise a technical system of production in a way which permits the direction of human labour and technology fitted into this system. The criticisms of this form of analysis have already been mentioned and can be briefly summarised. First, it has been suggested that Braverman's notion of deskilling is too monolithic; it asserts that managers are always trying to deskill workers, are always successful, and that there are no counter-tendencies. Second, Braverman's reference point for deskilling is a romanticised account of nineteenth-century engineering craft skill. This is historically limited and cannot be applied successfully in other areas. Third, he neglects the objective need for managers to utilise the co-operative character of labour in a production process. This results in a failure to acknowledge the alternative ways in which that co-operation can be achieved, and their potential effects on work organisation.[1]

To elaborate on these points, the empirical work on labour-process change carried out since Braverman's book has shown that Taylorism was a more complex phenomenon than Braverman supposed. Its development and diffusion was not as linear as presented by Braverman. In fact it encountered several obstacles including resistance from supervisors as well as workers. Furthermore there have been, and may still be, circumstances in which Taylorism may be 'reversed' and fragmented jobs recombined, in response to specific pressures by some or all of the participants in the historical evolution of particular patterns of work organisation. More fundamentally, it has been cogently argued that job fragmentation and job recombination *as management strategies* are not the coherent, planned strategies for control and production optimisation which the labour-process literature has suggested. This argument criticises the over-emphasis on foresight amongst entrepreneurs and managers, and the objective efficacy of organisational structures as means of achieving strategies even if they were to exist. A feature of both these criticisms is the emphasis on the capacity of workers and their organisations to respond to proposed changes in work organisation and in so doing force a compromise or bargain which makes the outcome diverge from any management strategies.

Despite these criticisms of the precise way in which specific work-organisational forms have developed, there has been little disagreement that Braverman's emphasis on deskilling is particularly relevant to an understanding of the period during which Fordism has been dominant. Some writers have suggested that these forms of work organisation have diffused into non-manufacturing sectors too. Studies of clerical work and computer programming show how such techniques have been used in an attempt to emulate the productiveness of mass-production organisation in other non-mass-production industries.[2] Indeed, taking car assembly as the paradigm for changes in the labour process, many analyses see a common direction of changes in all labour processes, expressed as a tendential movement towards Fordism which is seen simply as a labour process consisting of highly differentiated jobs with mechanised pacing. So, for example, computerisation of information processing is interpreted as a means of institutionalising mechanised pacing in other industries and creating 'assembly line' conditions in offices (Barker and Downing, 1980, p. 86).

However, there are several trends in job design and work organisation which run against such a tendency. There have been moves towards 'job enlargement' (the grouping of several similar tasks to form a job rather than specialisation in one or two of these), 'job enrichment' (adding higher order tasks) and 'group working' (where a large group of tasks is shared between a group of workers who allocate the tasks amongst themselves). These trends and experiments have been interpreted as the *reversal* of the trend towards an acute division of labour.

The reversal of the division of labour which is favoured by theories and experiments in job design is hardly surprising. Many of the earliest post-war examples of work reorganisation, such as job rotation, were specifically advocated as a response to the boredom which was felt by workers carrying out the repetitive operations of semi-mechanised processes. From this crude model relating repetition and boredom in work, industrial psychologists and sociologists have built up successive generations of theories linking aspects of work to psychological variables attributed to workers, such as motivation, satisfaction and the need for 'growth'.[3] These theories underpin attempts by managements to alter the allocation of tasks between various jobs in a way which reduces the fineness of the detailed division of labour.

Despite the fact that many of the experiments and descriptions of the field focus on social psychological variables as central to work organisation and job design activities, it is impossible to assess adequately the effect of these techniques at such a level. This is because of the pragmatic nature of most of the changes. The multiplicity of behavioural and organisational variables against which changes are assessed make general conclusions difficult to draw. However, there are indications that more significant conclusions can be drawn on another level. For example, in a European survey of the objectives of firms carrying out such changes, Wild and Birchall, (1975) found a wide variety of reasons given. Yet they concluded that a minority of exercises have been carried out with the primary intention of improving the behavioural characteristics of jobs. Where this was the case, it was usually as a means to an end, such as reducing absenteeism or labour turnover or improving product quality.

Many of the work-design experiments took place in assembly segments of mass-production manufacturing. One reason for this is the flexibility in work organisation which labour-intensive and relatively unmechanised segments, such as assembly, possess. Another is the low level of job control which the (typically women) workers involved in such work display. However, a more important implication of such management practices is the evidence in provides for the economic limitations which Fordist assembly lines display – even in mass-production industries. The constraints of Fordism have been relaxed to some extent by the introduction of flexibility into assembly operations through work reorganisation. Rather than having one assembly line for the production of increasingly differentiated products, firms have found that it may make economic sense to shorten assembly lines and even introduce *unit* assembly in some cases, since the costs associated with product change-over grow as a result of intensified competition in more slowly growing or even contracting markets (Kelly, 1982b, p. 14; Coriat, 1980, p. 42).

Such a strategy is compatible with the standardisation and modularisation of component parts in the more rigid, highly mechanised production lines. Modularisation is an attempt, by creating standardised sub-components, to allow mass-production techniques to be applied to intermediate products, increasing in volume and rationalising the variety of such products. This is, in fact, one attempt to overcome the 'productivity dilemma' highlighted by Abernathy (1978) where increased economies of scale are accompanied by decreased flexibility within mature industries. A combination of modularisation and flexible assembly procedures can act to reduce the rigidities of increased mechanisation.

A second indication of the limitations which Fordist forms of work organisation exhibit comes from examinations of highly capitalised industries such as food-processing and chemicals. In these cases, the work tasks are not the repetitive tasks associated in engineering with feeding machines or manipulating parts in a regular, short-cycle sequence. Instead they are less regular monitoring and interventionist tasks. The stochastic nature of these tasks, as we shall see, can pose difficulties in their individual allocation, contrary to what Taylorism demands. Under such circumstances, the introduction of flexibility in the allocation of labour becomes economically advantageous. Thus experiments in group working have been frequent in these industries, such as in paper-making, food-processing, oil-refining and chemicals-manufacture.[4]

Many writers on the labour process have dismissed these experiments in new forms of work organisation as minor or inconsequential, rather than see them as genuine difficulties within labour-process analyses.[5] The target has been to refute the liberatory rhetoric justifying the experiments by pointing out the limited nature of the benefits to workers which are said to result from such reorganisations. Whilst the recent experiments may be seen as more substantial than the 1930s 'human relations' techniques (contemptuously dismissed by Braverman as practised by the preventative 'maintenance crew' for the human machinery) they are nevertheless criticised as still hiding traditional methods of control. This view points to individualised responsibility for product quality and the linking of raised group output norms to bonus payments as the reality behind the rhetoric about increased autonomy and enriched work experiences.

Other authors have examied work redesign experiments more systematically, pursuing the suspicion that they represent something more than a new set of clothes for old principles. In a painstaking survey of published accounts of work redesign experiments, along with several case studies, Kelly (1978, 1979, 1982a) outlines three categories of work redesign:

- *vertical role integration:* where certain types of fractional work roles are reassigned to operators (for example, inspection, testing, scheduling, elementary maintenance). These roles would previously have been the responsibility of higher strata of workers (supervisory, technical or office personnel).
- *reorganisation of flow-lines:* where individualisation or a reduction of sequential work interdependencies takes place. This category mainly refers to the shortening of flow-lines, or a move to unit assembly mentioned previously. Often such a change is seen as a solution to the problem of changeover costs, delays caused by 'balancing' the flow of work along the line or attempts to avoid collective worker resistance to output norms.
- *flexible work groups:* here a group of tasks is allocated to a group of workers. Although this breaks down the permanent allocation of work tasks to particular jobs, it does not necessarily imply a radical despecialisation of function. This is because flexible allocation of tasks may be restricted to periodic inequalities in work loads after which workers may return to their specialised roles.

Kelly's analysis hangs on an examination of the claim made by job redesigners to have abandoned Taylorism as an organising principle of work. He argues that this claim is based on an identification of Taylorism with labour specialisation, whereas the latter was neither an integral nor a necessary feature of Scientific Management since, empirically, Taylor despecialised some groups of workers in his reorganisation of work methods. Taylor's Scientific Management, Kelly argues, was concerned with intensifying the utilisation of labour, along with other methods of increasing productivity, by a strategic use of individualised payment levels, training and work roles rather than specialisation (division of labour) as such, which pre-dated Taylorism itself.

So, to what extent can these forms of work redesign be seen as different to Taylorism? Kelly's view is as follows:

- *vertical role integration* is compatible with Taylorist techniques of assigning maximum work loads to a category of labour that is as cheap as possible. It only becomes novel when fractional supervisory and management functions are added to these jobs – it is then not compatible with Taylorism.

- *reorganisation of flow lines,* in contrast to vertical role integration is, in fact, a most thorough and consistent application of Taylorism. Individualisation of work roles, breaking up work teams, increasing the visibility of each worker's products, and the frequent use of individual pay incentives which follows the reduction of worker interdependencies – these are all techniques associated with Taylorism despite the despecialisation involved in such 'horizontal' role integration.

- the major innovation lies in the area of *flexible work groups.* This forms the distinctive practice of the forms of work organisation associated with the school of industrial sociology known as *socio-technics.*[6] The espousal of flexible work groups is a result of abandoning certain features of Taylorism; in particular, individual allocation of work, payment and accountability. But socio-technics has only done so, Kelly notes, to the extent that it has encountered limiting conditions of product or process uncertainty beyond which Taylorist principles have become less effective in achieving their stated goals. Socio-technical theorists have recognised the limits of their own work, but they have conceptualised it in terms of large task size rather than product or process variability.

Kelly's analysis has advantages over other, more descriptive reviews of job redesign experiments which results from its more refined model of work organisation.[7] The analysis does not see Fordism as a distinct type of production process however. Given the central role played by Fordism in our analysis, we shall reinterpret Kelly's findings in the context of our discussion of Fordism. We are assisted in this by Kelly's conceptualisation of the design of jobs in terms of work *roles:*

> In a service or production organisation, the main products and materials flow sequentially through a series of (more or less) interdependent work roles for processing. Attached to this *horizontal* organisation of roles are a number of off-shoots or *vertically* organised roles responsible for intervention in, or receipts from, the major flow of work. These vertically organised sections are responsible for such functions as maintenance, repair, material supply and collection, cleaning, inspection and supervision. It should be noted, however, that some of these functions may be designed into the main flow of work, such as brief quality checks, and the distinction between the two sets of roles is not absolute. (Kelly, 1979, p. 234)

We have seen how Kelly's threefold categorisation of work redesign is constructed from this model of work roles by specifying the addition of vertical, horizontal, and horizontal *and* vertical roles to an existing job. These correspond to vertical role enlargement, reorganisation of flowlines, and flexible working groups, respectively.

This model of production organisation – a sequence of interdependent horizontal roles within the sequential flow of materials – is in fact the division of labour associated with Fordism, whether it be a less mechanised assembly line, a more mechanised production line, or a highly mechanised flow process. Furthermore, the distinction between horizontal and vertical work-roles maps on to the contrast between the deskilled manipulating, machine-tending and feeding or material-transferring tasks assigned to workers 'on the line' and the technical, organisational and supervisory tasks of the vertical work-roles which are less machine-paced.

Another area of industrial sociology – organisation theory – has concerned itself with developing a model such as this. In spite of the reductionist nature of this work (Mortensen, 1979) some of its writers have examined the conditions under which organisational changes occur in firms, which have interesting implications for the structure of work role allocation. The dual structure of horizontal and vertical work roles is noted by Thompson (1967) in his discussion of the internal structure of organisations. Thompson describes organisations as consisting of a central core process surrounded by 'buffering' mechanisms:

> To maximise productivity of a manufacturing technology, the technical core must be able to operate as if the market will absorb the single kind of product at a continuous rate, and as if inputs flowed continuously, at a steady rate and with specified quality. Conceivably both sets of conditions could occur; realistically they do not. But organisations reveal a variety of devices for approximating these 'as if' assumptions, with input and output components meeting fluctuating environments and converting them into steady conditions for the technological core. (Thompson, 1967, p. 20)

Buffering, then, takes place on the 'input' side of the core process by ensuring the continuous availability of raw materials often in a varying market and feeding them steadily into the production process, and on the 'output' side by sales and marketing activities which try to ensure a constant or rising level of sales. The latter is especially important where goods are perishable or expensive to store. But it is in the manufacturing system itself that the distinction between core process and buffering mechanisms is of interest to us, for these define the difference between the two types of work-role.

Workers in the core process will be allocated tasks which are highly prescribed (or as Hales, 1980, puts it, 'preconceptualised') in relation to the technical core. Hence, where labour is an integral part of the core – for example, in assembly or machine feeding or unloading – attempts are made to specify these tasks in a mechanical way. Therein lies the importance of time and motion studies and ergonomics, in fitting humans to machines. In contrast workers in the buffering functions are more concerned with ensuring the *continuity* of production, than with contributing directly to the core process. This peripheral work, then, is more contingent and less directly paced. These functions of maintenance, scheduling, allocation and supervision of labour, quality control and testing correspond to the vertical work-roles described by Kelly.

The separation of core and peripheral work (horizontal and vertical work roles) is one of the basic features of Fordist labour processes. It is the relationship between core work, and its specification by peripheral workers which is at the root of the separation of conception and execution (Braverman), preconceptualisation (Hales) or forms of technical control (Edwards) in Fordism.[8]

Having established the distinction between core and periphery processes, organisational theorists go on to describe the effects of changing environments on management structure.[9] Our concern is with the changes that take place in work organisation and technology in response to such environmental change. The important point here is that the Fordist division of labour in production depends not just on scale, but also on stability. We have argued that Fordist processes gained their economic viability under certain conditions: growing markets for regular commodities ensuring continuous flow of work with adequate supplies of labour and raw materials. Where these conditions do *not* apply – where there is variability in raw materials or work tasks, where technical change is high or products vary (as in small-batch production), where markets stagnate and raw materials costs increase steeply causing rapid changes in demand (as in the car industry since the 1960s) – the ability to achieve a stable core of horizontal work roles is difficult.

Kelly's three forms of work redesign (vertical role integration, flow-line reorganisation and flexible work groups) would seem to have similar potentialities for increasing the variety of work but there are important differences in the structural basis of these changes. Vertical role integration, given its minor part in reducing horizontal task interdependencies is important as a means of using the 'idle time' (porosity) of particular jobs. This is achieved by allocating the worker certain peripheral tasks which need not be performed at any specific time. Since they are not sequentially dependent on previous and subsequent tasks, they can be performed in between regular tasks.

Reorganisation of flow-lines introduces a different type of flexibility into a Fordist process. The move from sequential to parallel processing entails the circumvention of certain Fordist principles since it allows different workers to produce at different rates without problems of bottlenecks and line-balancing, as well as to produce different products, a feature especially important in small-batch production. Kelly found the majority of examples of reorganisation of flow-lines occurred in assembly or sequentially organised work in white-collar work areas. The resulting individualisation of tasks permitted management to increase working-time through reducing the amount of time spent in handling and waiting for parts. Quality improvements were also possible given that individuals were identifiable as responsible for the production of particular items.[10] In the third category of work redesign, that of flexible groups, we noted Kelly's explanation of this as a response to the problem of allocating labour economically under conditions of product or process uncertainty. He is able to support this position by reinterpreting the group-work experiments of key socio-technical writers as ultimately based on the intensification of labour. Yet the forms of flexibility which group work offers over Fordist configurations extend further than the quantitative allocation of labour. Group working may also involve multiskilling and planning functions (as we shall see in the case of Group Technology, discussed in Chapter 6). Furthermore, this form of organisation is a potential threat to forms of job control which are built upon demarcation and skill structures (Hull, 1978).

In summary, then, the significance of the work-redesign experiments of the last two decades is that they try to break the structure of jobs organised around the division between horizontal and vertical work roles which is characteristic of Fordism. The reintegration of such roles will be used later as one feature of a definition of neo-Fordism. However, just as we have defined Fordism using a cluster of variables encompassing standard products, mass markets, dedicated production technologies, homogenised consumption norms and particular labour processes, we can define neo-Fordism using a similar range of variables. We have already noted some of these: a more differentiated range of products corresponding to increased heterogeneity of consumption and a more complex, differentiated market with shifting levels of technical change as constituting an important part of capitalist competition.

In the next section we shall concentrate on the *technological* elements of the transition from Fordism to neo-Fordism in an attempt to relate the technological and organisational dimensions of the labour process visible during this transitional period.

NEW TECHNOLOGIES AND FORMS OF MECHANISATION

Whereas work redesign theorists attempt to construct criteria by which jobs *ought* to be constructed for psychological ends, their claim is also to reconcile these criteria with economic ones.[11] What is missing from these analyses is an acknowledgement of the origin of these constituent tasks. Work redesign takes a contingent and partial view of technological change, orientated towards the bundling of tasks into jobs only *after* the technology has thrown up the tasks to be 'sorted'. As Herbst (1974) has pointed out, despite its theoretical claims to the contrary, even socio-technics has taken the technical system for granted in its actual design of work, concentrating on manipulating the social system around pre-set variables.

What we have emphasised, however, is that we must recognise technological change as a significant factor in the changes in labour processes. Attention can thereby be drawn to the 'creation' of technologies and to the social relations implicit in the design of machinery. This approach contrasts sharply with the uncritical, emancipatory view of technological change put forward in the study of 'alienation' and 'automation' by such writers as Blauner (1964). In the recent debate on the effects of micro-electronics on the future of work, a similarly positive view of production technologies based on the silicon chip inherently leading to upgraded work is widespread.[12] Often this is the result of an assumption that core processes in all the major industries will be swiftly mechanised, resulting in the more skilled and potentially satisfying vertical work-roles remaining.

The drawback to such general approaches to technological change as upgrading or degrading work is that they attempt to generalise across a large number of technical changes in a large number of industries. But we are not saying that a contingent approach to technological change is required which must examine each case independently. Obviously, there must be some contingency in any model of the role of technological change in work organisation developments to account for differences in such things as the effects of management style or of workers' collective strength. Yet such factors will only be operative within the limits created by general trends and particular forms of mechanisation. It is these issues which we wish to discuss in this section.

We have discussed in Chapter 2 the conceptualisation of 'automation' by Bright and Bell. We shall now develop Bell's ideas further in order to clarify the differences in work organisation between Fordism and neo-Fordism. In contrast to Bright's unilinear and progressive succession of stages in mechanisation, Bell's three dimensional model of 'automation' is constructed in such a way as to stress the relative independence of the different types of mechanisation. This overcomes a problem of Bright's scheme, which is the arbitrary nature of the sequence of his levels of mechanisation.[13]

We may envisage, therefore, seemingly highly 'automated' machines whose degree of mechanisation is in one particular dimension only. Examples of this situation might be a very precise milling or grinding machine which was hand-fed with material and required manual setting up (that is, having a highly mechanised transformation function whilst the transfer and control functions are relatively unmechanised); or a manual assembly line fed by a conveyor-belt would represent a situation of mechanised transfer, whilst the transformation of parts (assembly) would be mechanised to a low degree in this case, perhaps using hand or pneumatic tools.

In attempting to link together technological change and work tasks the most successful approach seems to be to analyse the latter in terms of the residual functions left unmechanised. That is, workers in the core process perform the tasks that machinery is unable to perform. Bell's work is particularly useful in that regard. Using his model we might represent the residual tasks for the human operator under different conditions of mechanisation as in Figure 5.1.

As each dimension becomes increasingly mechanised (represented by the bold lines) the corresponding human tasks are reduced (represented by the shaded areas). The configuration in *A* represents non-mechanised work, such as 'craft work', that is, the use of hand-tools with work in progress moved by hand, and perhaps consisting of tasks which rely highly on skill, planning and design. The introduction of machines using non-human labour power corresponds to *B*. When machines can be linked together by mechanised transfer systems, we have *C,* corresponding to the transfer line. Yet the forms of labour-intensive assembly lines usually associated with mass production work are nearer to *D* since they commonly only use hand- or powered-tools (spanners, screwdrivers, sewing-machines, soldering irons). The highest stage in this Fordist phase, *E,* is characterised by self-acting control systems (though still with some manual control), until recently mainly found only in flow-processes, such as oil, paper, steel and chemicals and, sometimes, food-processing.

As far as the organisation of the labour process is concerned we can identify particular forms of work organisation generally associated with these different cases of mechanisation. Thus the largely unmechanised case, *A,* is associated with both craftwork *and* unmechanised industries characterised by a high division of labour (for example, much office work). The use of machinery in the transformative elements of production is represented by *B*. This creates tasks of machine-minding and maintenance besides the transfer tasks (feeding raw materials and removing finished products and waste, as well as the routeing of products between machines) and

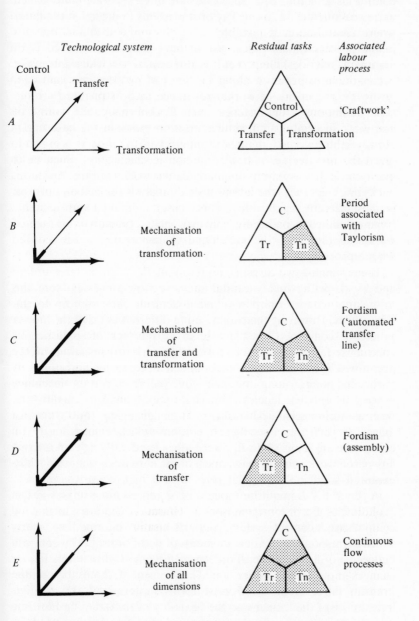

FIGURE 5.1 *Residual tasks in the development of mechanisation*

control tasks (setting up). Such tasks are often separated into skilled and non-skilled tasks, using Taylorist methods to prescribe the latter where routinisation is possible. This division is a crucial one for labour-process theory since, as we have seen, it is the basis of arguments over deskilling. Yet it is also central to a wider analysis of occupational segregation along the lines of gender, race and age, given the propensity of worker-resistance to be structured around skill definitions (for instance, see Cockburn's, 1983 study of craftwork in printing). A similar position exists in the case of *D*. However, in this case, the labour intensity of the process is evident, given the low degree of transformation mechanisation. Such tasks often consist of assembly or manipulative work (testing, finishing, packaging). As such, the labour process analysis centred on Fordism is most directly applicable to this case; a high labour content, fragmented work paced by conveyor belts, repetitive, routinised work – all these are the archetypal characteristics of the Fordist labour process.

The remaining two cases: *C*, representing the 'Detroit' transfer line and *E,* depicting highly capital-intensive flow-processes were the most 'automated' examples of manufacturing processes up to the early 1970s. However, the two are quite different as far as the labour process is concerned. In the case of *C*, the lack of sophisticated control systems makes the forms of work organisation highly prescribed. Thus there are tasks which need to be performed at particular points along the line – overseeing a certain machine, making up for deficiencies in mechanisation, checking quality – as well as non-mechanised tasks such as changing tools, loading magazines and certain other operations which cannot easily or cheaply be mechanised. In *E*, on the other hand, although some fixed and paced work can still exist, much of the core work is mechanically executed leaving the vertical roles with a high control content.[14]

In general *A-E* might be thought of as representing phases in the evolution of Fordist mechanisation. However, the introduction of computerised control systems permits certain intermediate states which correspond to the developments of neo-Fordism. The contrast between Fordism and neo-Fordism at a technological level, lies in changes in the control dimension. The forms of mechanisation under Fordism differ from neo-Fordist forms in terms of the relative inflexibility of the former and the flexibility of the latter. In Fordism incremental changes in the control mechanisms incorporated in machinery have the effect of reducing the variety of operations possible with a single piece of machinery. The creation of specialised machinery, or 'dedication' of otherwise flexible, tool-like machinery using such things as fixtures and stops, became feasible only with increased volumes of production. But this was only part of the process, the integration of a collection of such machines into a higher stage of this type of mechanisation – the transfer line – further exacerbated such inflexibility (see Abernathy, 1978). The creation of Fordist, highly mechanised systems of production, then, was feasible only under certain highly specific conditions, a point we have argued earlier, in Chapter 2.

What neo-Fordism is about at a technological level is a development and refinement in control systems both on individual machines and in the co-ordination of sets of transformation and transfer mechanisms. The next chapter will detail this for the small-batch engineering industries by examining the development and use of numerically-controlled and computer-controlled machine tools in cellular systems and, ultimately, linked together in so-called computer-aided manufacturing systems.

However, neo-Fordist developments in technology need not take the form of integrated, highly mechanised processes of the 'automated factory' model. Intermediate stages of neo-Fordist mechanisation corresponding to cases *B* and *D* can occur. The existence of stand-alone computer-controlled machinery (machine tools, sewing machinery, weaving looms, etc.) is the neo-Fordist equivalent to *B*. Whereas the use of computer-controlled transfer machinery, such as in the robotic handling systems or flexible pallet transfer systems replacing the conveyor belt as used in Fiat's car assembly plant (Amin, 1982) corresponds to *D*. Such changes may themselves be transcended by further mechanisation, such as the use of mechanised handling between discrete machines, say, by robots, in the former case or by the mechanisation of assembly in the latter. Of course, such changes will not necessarily occur in a step-wise fashion mirroring the logic of this categorisation. It is quite likely that significant changes in the *product* would be a prerequisite to further mechanisation (Barron and Curnow, 1979, p. 131).

It is important to emphasise that the distinction between Fordism and neo-Fordism is being used to distinguish between the dominant forms of production associated with leading industries and sectors in different phases of economic growth. It might be argued though, that this distinction is weakened by the existence of flow-process plant which exhibits a high degree of mechanisation along *all* Bell's dimensions, as well as displaying some elements of neo-Fordist work organisation. In fact, the early studies of Blauner (1964) and Woodward (1965) both identify process plant as a distinctive form of 'production technology'. Blauner, however, suggested that chemical plant technology represented the eventual direction of development of the less mechanised, batch and mass production systems for discrete items. It was not possible for him to envisage such a development in terms of the crude mechanical, electro-mechanical and mainframe computer-based control systems then available, around which the 'automation debate' of mid-1960s revolved.

Nevertheless, Blauner's study was ultimately limited by his overriding concern with the fate of blue-collar work. By treating 'automation' as the mechanisation of horizontal work roles, leaving the upgraded vertical roles intact, he was able to infer a correlation between increased 'automation' and work characterised by decreased 'alienation'. In concentrating on shop-floor work he neglected the major area for the impact of 'automation' – the conceptual areas. The vertical work-roles of design, planning, scheduling, etc., which Blauner took as remaining fairly constant, are in fact the areas being reconstituted by information technologies and neo-Fordism. Hence, to argue that process plant exhibited neo-Fordism in only valid inasmuch as *part* of its work-role structure displayed particular aspects of neo-Fordism. With information technologies and increasing computerisation of process control, the residual *vertical* tasks are also likely to be greatly transformed.

A similar phenomenon is associated with neo-Fordist technological changes in other sectors. For example, computer-aided design (CAD) is often twinned with computer-aided manufacturing (CAM) as important developments in the potential integration of conceptual and executive labour processes via linked CAD/CAM computer systems. Information-technology-based systems for stock control, production planning and scheduling are essential parts of the informational infrastructure around which neo-Fordist labour processes are structured.[15]

NEO-FORDISM

In summary, our definition of neo-Fordism consists of several interdependent strands. These are:

• *technological* developments in control mechanisation which weaken the link between mechanisation and scale. This 'flexible' form of mechanisation permits increased variablity in products and processes to be accommodated at higher levels of mechanisation.

- *organisational* changes which tend to favour the adoption of work roles, rather than individual repetitive tasks organised on an hierarchical basis.

In contrast to a core or line process around which work is organised, there is more potential for a functional division of labour structured around semi-autonomous groups of workers and machinery whose co-ordination depends on the setting up of:

- *informational* infrastructures which integrate different productive sub-units (semi-autonomous groups, machining centres or different interdependent, geographically-separated production units) with control of materials flow, stock control and production planning.

Other writers concerned with neo-Fordism have used the term in three senses:

- to refer to the intensification which occurs by means of 'thinning out' of hierarchically-structured work-organisation. The integration of previously distinct tasks (job enlargement and enrichment) is seen as a development of Fordism in response to changed conditions of labour market supply and problems of the control of labour (for example, Palloix, 1976).
- changes in the organisation of working practice which can accompany increased mechanisation. The development of semi-automatic machinery which does not require the continuous presence of a human operator and/or where tasks are variable in timing and content (for example, process plant) may preclude the distribution of repetitive tasks on an individual basis. New forms of work organisation, other than those associated with Taylorist and Fordist models are necessary if labour is to be used economically (Kelly, 1978, 1982a).
- the extension and development of mass-production techniques to new areas of production and services previously outside the ambit of such techniques. This includes attempts to rationalise and 'automate' small- and medium-batch engineering through the use of Group Technology and Flexible Manufacturing Systems. Similarly, the use of information technology in clerical and service

industries would be potential fields for the development of neo-Fordist techniques (see Aglietta, 1979).

Our use of the term embraces all of these. Most labour process theorists tend to analyse the significance of neo-Fordism only in terms of the attempt by capital to intensify work. Even though some of them draw attention to certain aspects of the other two, the focus in such cases is predominantly on new modes of control or increased exploitation of labour.[16] There is something to say for such an analysis. Many firms, in their attempts to ease the squeeze on profits arising from increased competition, have used job redesign techniques as a means of effectively increasing the work-load. Indeed job redesign consultants have played on such productivity-enhancing potential in 'Quality of Working Life' changes as a means of selling work humanisation schemes to hesitant managers.

It would be unwise, however, to write off work redesign as just management's latest fad for intensifying work. It does also focus attention on the human aspects of production system design and challenges aspects of current engineering and systems design practice.[17] In these areas, if not in the sphere of economics directly, the prevalent orthodoxy of treating labour inputs as similar to, or subordinate to, fixed capital inputs is being contested.

However, job redesign as a set of developing techniques is more than just a productivity tool. There has been a tendency amongst writers of the labour process to extrapolate their analysis of deskilling as the progressive routinisation of tasks to encompass, in time, *all* work. What this extrapolation overlooks, however, are the contradictory elements in such a process. Earlier, we considered how such contradictions can form the basis for job control by workers including the maintenance of craft-type skills. This applies particularly to work involving a degree of variety where routinisation is not economical or where successful worker-resistance has avoided it. If we take the case of workers whose jobs have already been 'deskilled', in the sense that they could not lay claim to craft-type job controls, a deskilling analysis offers little in the way of restitutive strategies to regain non-degraded work. We have argued that this is a reflection of the individualistic nature of a deskilling analysis based on the notion of craft skills. What is needed is an analysis which recognises the *collective* nature of modern production processes. Job redesign analysts have made at least exploratory movements in this direction. The élitism of protecting the craft skills of 'labour aristocrats' from degradation should be replaced by a consideration of what strategies may be appropriate to semi-skilled and unskilled workers who form the majority of manufacturing workers, most particularly the majority of women and black workers. In this sense, the concept of neo-Fordism has particular advantages – not least the potential upgrading of work which it may entail. Furthermore, given the imminent changes in the organisation of service occupations, such an analysis is much more likely to be appropriate to these workers than one based on an over-humanised conception of a craft worker.

In this chapter, we have tried to develop such an analysis by making clear the distinction between Fordism and neo-Fordism. The conceptual framework which has been developed will allow us to operationalise the concept of neo-Fordism by investigating changes currently taking place in two sectors: small-batch engineering and services. These have been chosen for their potential importance amongst the leading sectors in any future economic upswing.

6 Neo-Fordism in Small-batch Engineering

INTRODUCTION

In this chapter we will take the example of small-batch engineering and argue that this is a sector which is undergoing neo-Fordist changes. The sector is important because of its position in the production of capital goods. Hence, transformations in technology and organisation within this sector which result in cheapening its products are likely to have widespread effects throughout many other sectors which use these capital goods.

In discussing the trajectory of Fordism, it was clear that the pre-conditions for the diffusion of Fordism in this industry were not satisfied, given the range of products and instability of markets characteristic of batch engineering production. Thus the major part of mechanical engineering was relatively isolated from changes associated with Fordism, permitting the continued existence of skilled work, organised around craft structures as the central core of small-batch engineering labour processes. New mass-production areas of engineering grew up alongside the traditional industry.

Yet technological changes and organisational innovations which prove more appropriate to the transformation of small-batch engineering have been attempted. As we investigate such changes below, we will argue that such developments do not represent a mere extension of Fordism into small-batch engineering with somewhat limited success. They entail a *discontinuity* with Fordism and are more akin to the configuration of neo-Fordism. At the level of the labour process, the development of sophisticated control mechanisation and production organisation which is *not* based on a flowline core production process, makes orthodox labour process analyses problematic. The separation between programming and operation of 'Numerically Controlled' (NC) machinery, which Braverman noted, is uneven and is often later reversed on 'Computer Numerical

Control' (CNC) machines. This can not merely be explained by worker resistance and managerial strategies on the shop-floor. Nor can such 'variability' in the division of labour be simply relegated to the contingent interplay between various wider factors, such as product markets and trade union organisation. The problem of explaining such variability lies in a confusion at the analytical level. Changes which prefigure neo-Fordism are difficult to analyse on the basis of Fordism.

What follows, therefore, is a description of the neo-Fordist developments within small-batch engineering. Obviously, this is a continuing process and is by no means complete. Nevertheless, our aim is to demonstrate the value of such an analysis in accounting for current changes in this sector, rather than one of the more or less successful playing out of Fordist tendencies. After an initial description of the sector, we will examine some existing labour process analyses of technical changes in small-batch engineering and contrast these accounts with a more detailed description and analysis of successive changes in the production process which reflect the contradictory goals and partial solutions accompanying the process of transformation in this sector.

SMALL-BATCH ENGINEERING

Despite the preponderance of mass-production techniques in accounting for the bulk of the output of the engineering industry, the majority of *employment* within mechanical engineering involves production in batches rather than mass production. In Britain, mechanical engineering accounts for approximately one third of total engineering employment (Swords-Isherwood and Senker, 1980). Hence radical changes in batch engineering will have an appreciable effect within the engineering industry as a whole. This is in addition to the 'knock on' effect on other industries using capital goods made in the mechanical engineering industry. In Chapter 4 we noted that the cheapening of capital goods production was particularly import- ant for a renewed sustained upswing in economic growth; so transformations within small-batch engineering production are of strategic significance for all sectors of production, though it is worth noting that the techniques exemplified in this transformation are also applicable to other, non-engineering, batch-manufacturing sectors such as clothing, footwear or furniture and might indeed be crucial if these sectors are to recover from their current depression.

In the mechanical engineering industry despite the wide range of separate processes carried out (forming, casting, welding, cutting, surface finishing, making tools, jigs and dies, and assembly) it is metal-cutting which is the major activity. Although there are new techniques which are likely to be important in other processes, the largest impact on organisation, employment and skill levels as well as on the cheapening of products is most likely to result from technological and organisational changes around the cutting of metals. Significantly, most attention has been addressed to the effects of computerising control functions in metal-cutting machinery.

In a later section, we will extend this focus to cover developments in the mechanisation of parts-transfer between machines as well as computer co-ordination systems (and information systems) on which mechanisation of transfer would need to rely. For the moment, however, the discussion will be restricted to the debate over the labour-process implications of numerical control in metal-cutting machine tools. The discussion will be broadened subsequently by looking at the organisational changes associated with 'Group Technology' and their relation to technological changes before leading on to an interpretation of current transformations in small-batch engineering which develops the concept of neo-Fordism.

NUMERICAL CONTROL AND THE CONTROL OF WORKERS

In the debate over 'deskilling', one of the commonest examples has been that of changes in metal-cutting machinery particularly the introduction of Numerical Control (NC). NC lathes, for instance, replace the manual controls of cutting speed and tool movement, which rely on work planning and the hand–eye co-ordination of skilled machinists, with mechanised controls operated by a sequence of instructions from a punched paper tape, or directly from a computer. Braverman's work contained an extended argument (1974: pp. 184–248) on the separation between programming and machine operating. He used this as an example of the division of labour between conception and execution which, for him, represented the hallmark of changes in the labour process under monopoly capitalism. By removing programming functions to the office, capital was able to increase control over the labour process, undermining the skills of the skilled worker, and to use this control to increase the rate of production, whilst at the same time using cheaper unskilled machine-minders.

Braverman's work has been extended in research conducted by Noble (1979). He goes further than Braverman in tracing the historical development of NC machinery in preference to other metal-cutting technologies which did not permit deskilling to be imposed on skilled machinists. Nobel notes how NC machinery, along with a highly sophisticated software package, was heavily underwritten and championed by the US Air Force, thus highlighting the crucial role of military requirements in the development and diffusion of the new technology. Whilst the high-technology nature of such capital-intensive equipment favoured the complex machining requirements of Air Force orders, it also gave a distinct advantage to the large firms in bidding for orders, since with Air Force support they could afford to purchase and use such equipment more than small firms. This further aided the process of monopolisation in the industry and was an important element of the prime contractor/ subcontractor division amongst the US aerospace industry noted by Jones (1982a).

Yet the main weight of Noble's argument is put on the contestation of control on the shop-floor. NC is advantageous to management because of the potential which (they think) it holds for 'enhancing their authority over production, despite questionable cost-effectiveness' (Noble, 1979: p. 34). The technology has such possibilities for increasing managerial control that it is 'seized upon' by management notes Noble, quoting Braverman. Yet by accentuating capital–labour relations more than other important relations – capital–capital (competition), labour–labour (gender and skill), state–capital (the 'military industrial complex') – besides the technical capabilities of NC over rival technologies, Noble reduces his account to playing out a political battle for control on the shop-floor. What is more, by elucidating the difficulties of estimating the economic returns and therefore the viability of new technology, Noble (1979, p. 35) closes off economic explanations for the use of NC on the basis of the existence of economic uncertainty.

Having relegated economic considerations to the level of justifications on the part of management, Noble is able to continue to describe the politics of shop-floor conflict for control in terms of the dialectic between management 'dreams' and shop-floor 'reality'. Manufacturing engineers and managers develop technologies and organisational systems to subordinate and displace workers: yet, the reality encompasses a failure to achieve such dreams. Stoppages and strikes over rates of pay for the new machines and thus the assigning of labour grades are common, Noble argues, especially in plants where machinists' unions have had a long history. However, the limitations of the NC machinery itself also work against the technological illusion. The operator's role may be reduced in comparison with some of the craft-like skills of workers on the old machinery, but the yield and quality of the products are still heavily dependent on the 'unskilled' operatives. Hence new niches are revealed for worker control and management is once again confronted with problems of 'motivation'.

Another empirical study, this time a British one, is also less convinced as to the inevitability of the deskilling process as described by Braverman. Wilkinson (1981) investigates the politics of technical change in batch manufacturing, including machine shops. He also sees the eventual pattern of work organisation and division of labour as the outcome of a process of contestation and negotiation. The incremental nature of technical changes in batch processes allows, he argues, a greater degree of contestation over the organisation and introduction of work and machinery than is the case when whole systems are designed and proved outside the workplace and then installed *en bloc* (for example, flow-process plant). He is able to demonstrate this 'shop-floor politics' through various case studies.

What becomes clear from these studies is the continuity which exists between working practices on the old and new machinery. Wilkinson attributes this to the interests of various parties (management, programmers, operators) being played out in the bargaining arena during the process of introducing the new technology. For example, during the period when a particular piece of 'automatic' machinery was being 'debugged', there was an opportunity to establish certain working practices (for instance, using manual overrides to control the speed and quality of machining) which were subsequently retained. This was a way of 'clawing back' skills which were supposedly incorporated into the 'automatic' system. These emergent working-practices were found by Wilkinson to have an important role in determining the final form of work organisation. In fact, it was often the case that the quality of products crucially depended on the 'knowledgeable' intervention of operators despite the degree of automaticity claimed for the machinery (Wilkinson, 1981, p. 123).

As far as the practices of management was concerned, these

varied. There were several attempts to avoid losing control of the production process to the workforce by 'debugging' the new machinery away from the shop-floor. But direct control was not the only strategy used by managements. Wilkinson also found a case where management recognition of the importance of workers' skills and goodwill resulted in a system of improved working conditions and job rotation (Wilkinson, 1981, p. 198). This contributed to a manifest variability in the eventual system of work organisation between firms using similar technology. The general thrust of this work, then, is very much against a deterministic approach to skill and organisation changes. As Wilkinson argues:

> The contribution of the case studies to the skills debate is (i) that technology is not autonomous and therefore does not have impacts regardless of the social context, and (ii) that forces of competition do not impel firms to impose specific skill structures. Rather it was shown that the technical and social organisation of work – and thus *real* skills – is a negotiable phenomenon, determined by social and political processes whose outcomes were never very certain. (Wilkinson, 1981, p. 210)

There is a major problem associated with an analysis of this sort. The uncertainties which Wilkinson seeks to incorporate into his 'bargaining' model of labour process change are simultaneously used to hold off any explanations based on economic or 'efficiency' criteria. Thus, like Noble, the complexities of economic evaluations of new technologies *in practice* is interpreted as leave to put such considerations aside:

> In fact, what our case studies show is that arguments about the efficiency of new production technologies are often *no more than scientific glosses which conceal or obscure the political considerations* which have gone in to decisions on technical change and work organisation. (Wilkinson, 1981, pp. 194–5; emphasis original)

By concentrating on the legitimatory functions of claims to efficiency or economy, the political economy of technological change becomes reduced to a politics of shop-floor struggle over new technology. Just like many labour process studies there is a reification of 'control' which hides the relationship between the macro-level *structural* conditions of capital accumulation (markets, competition, product change) and the micro-level *interactional* arena of shop-floor negotiation in particular firms.

A further empirical study of NC which contests the deskilling thesis is that of Jones (1982a). He introduces additional factors which influence the deployment of skills in NC machining such as the nature of the products, the composition of demand, the supply of skilled labour and the structure of trade union organisation. This analysis is not incommensurate with the general position argued here – at least as far as the range of factors which relate to the structure of the labour process in NC machining is concerned. For example, he notes the tendency for less complex components and increasing batch sizes to be associated with NC operation by semi-skilled workers (Jones, 1982a, p. 193). Also he found that criteria for investment in NC machinery were less concerned with the cost of labour (*contra* Braverman) and more concerned with such things as the increased accuracy of NC machined components and the shortage of highly skilled labour in the local labour market.[1]

The central feature of these writers' work relates to their explanation of the observed variability of skill deployment in NC machine shops. Noble accounted for this in terms of the limitations of NC machinery and management control systems, which was expressed as the continued dependence of management on operators' tacit knowledge. The shop-floor battle for control continued, however, with management trying to deskill and control workers directly. Wilkinson also drew attention to this phenomenon. Yet, unlike Noble's assumption that an alternative division of labour reuniting programming and operating is only feasible on the basis of collective worker resistance (Noble, 1979, pp. 45–50), Wilkinson notes the success which some NC operators had in gaining access to machine-based editing facilities to the extent of using the latter to create their own programmes! Wilkinson has a less deterministic model of the direction of change of the labour process, acknowledging that different management strategies are possible other than an overriding concern with direct control. Furthermore he notes significant variations in the degree of constraint which factors such as the design of technology, and the result of the implementation and debugging procedures have on the labour process. This can lead to consequent variations in the eventual form of work organisation.

Jones' position, on the other hand, is much more ambivalent. The variability of skill distribution and work organisation which Jones observed is used as a means of rejecting any general or unilinear tendency towards deskilling or skill polarisation. The amount of variation amongst Jones's sample, accounted for by multiple and interacting factors of product and labour markets, trade union organisation and technology, leads him to a position of indeterminacy. He rejects the existence of any unitary tendency towards a particular structure of the labour process. He assumes that there is no reason why other areas of manufacturing besides small-batch engineering should not display an equivalent diversity of organisational forms (Jones, 1982a, p. 200).

Rejecting any over-determining direction of labour process development, such a conclusion, whilst reflecting the degree of complexity of labour process change in practice, ends up agnostic with respect to general dynamics of labour process change. It is a belief in total contingency. In other words, the inadequacy of a simple structural model of technical change under capitalism (that is, variants of 'deskilling') becomes the occasion for neglecting the construction of a more adequate structural analysis of contemporary changes in the labour process.

It should be clear that our argument does not seek to deny the empirically observed diversity of work organisation under NC. In that sense, the determinist overtones previously criticised in labour process theory are avoided. Nevertheless, the *structural* argument being proposed is based on a model of dominant configurations of the labour process (Fordism, neo-Fordism) being associated with certain industries in different phases of the long wave.

The reason why Wilkinson and Jones are able to reject a inherent 'deskilling' tendency on the basis of their case studies, is that they are looking for evidence of a transformation towards *Fordism* within small-batch engineering. With large batches of simple and similar components examples of Fordist labour processes can be found. Yet this accounts for a minority of cases. Hence the diversity. What should be examined is whether we can analyse the 'automation' of small-batch engineering as representing a shift from a 'craft' to a *neo*-Fordist labour process. At the end of Chapter 5 we defined the labour process of neo-Fordism as consisting of several interdependent strands – technological, organisational and informational. We think that contemporary changes in the labour process of small-batch engineering can be analysed in terms of these aspects of neo-Fordist transformation. In examining successive attempts to transform small-batch engineering we hope to demonstrate that neo-Fordism is a more useful concept with which to understand the dynamics of these structural changes than the labour process analyses.

As different aspects of neo-Fordist transformation are discussed it will become apparent that initiatives and experiments within the industry often focus on one or, sometimes, two of the three aspects; hence the history of such changes is one of partial solutions. The result, then, is more a residue of the most useful techniques than an accretion of the most successful experiments. Thus an organisational innovation, 'Group Technology', which is discussed below, becomes a basis of flexible manufacturing cells rather than being widely diffused itself, *in toto*, as a new way of organising machine shops. Similarly, NC machinery is limited in its potential for 'automating' skilled machinist work, not least by problems of ensuring quality of output (Doring and Salling, 1971). Hence deskilling may prove to have limited economic returns as labour costs become less important than scrap rates or rates of machine utilisation under conditions of rising capital intensity. NC machinery, then, may be more economically justifiable if operated by skilled rather than unskilled workers.

A limitation to any account of changes in small-batch engineering is the inevitable uncertainty of the final outcome given the fact that the process is still in train. Nevertheless, it will be argued that there does seem to be sufficient evidence to support a neo-Fordist explanation of current changes in the industry. It is in this context that shop-floor politics should be situated at the level of particular firms.

ORGANISATIONAL CHANGE IN MACHINE SHOPS: THE ESPOUSAL OF GROUP TECHNOLOGY

Since the last century the dominant form of plant layout in batch engineering has been that of *functional layout* in which machines are grouped by machine type in 'shops' or work area (see Figure 6.1(c)). This structure mirrors the range of individual processes (such as drilling, welding, turning, milling, etc.) which are available for use in manufacture. It also reflects the dominant forms of occupational specialisation based on skilled 'trades'. Depending on the product being made, the sequence of operations in manufacture, and thus the routeing of components or semi-completed products will vary. This necessitates some control over the transfer sequence of parts between shops, and also some means of monitoring the location of work in progress. At present this is done bureaucratically by means of production planners and progress chasers. It is only when products are standardised to a large extent, in terms of the sequence and type of manufacturing operations, that the *line layout* (Figure 6.1.(a)), characteristic of mass production, is viable.

(a) Line layout (Machines grouped by component family)
   [Component specialisation]

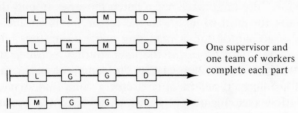

One supervisor and one team of workers complete each part

Machines in each line always used in the same sequence

(b) Group layout (Machines grouped by component family)
   [Component specialisation]

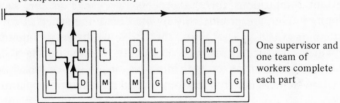

One supervisor and one team of workers complete each part

Machines in each group need not be used in the same sequence

(c) Functional layout (Machines grouped by type)
   [Process specialisation]

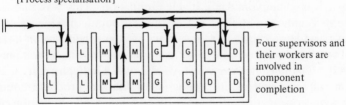

Four supervisors and their workers are involved in component completion

Key   L = lathe   M = miller   G = grinder   D = driller

FIGURE 6.1   *Types of factory layout*

SOURCE   *Adapted from Burbidge (1975).*

However, the price of maintaining flexibility by a functional layout is very high compared with that of mass production. Cook (1975) estimates the difference in cost between conventional batch manufacturing and mass production to be in the order of ten to thirty times as much per unit of product. In part, this is due to the cost of setting up the machinery between batches which can result in the machine only cutting for an average of 15–20 per cent of the time that the workpiece is on the machine table (Williamson, 1972, p. 142). Furthermore, the time taken for a batch to move around the plant can account for most of its life in the factory. Williamson suggests that this is accounted for in queueing and waiting time; this is in addition to the time taken for the long journey between different shops (Williamson, 1972, p. 142). The different paths taken by different products result in an extremely complicated pattern of material flow (see Figure 6.2(i)) which adds to the problems of production planning and results in lengthy 'throughput times' and hence high levels of work in progress.

The particular problems of batch-production organisation then, have important implications for the process of change in this sector. Attempts to increase productivity by increasing the speed of cutting has a minor effect on total costs; this is due to the small amount of time which a particular machine is actually working during a shift. Indeed the substitution of individual NC machines or 'machining centres' (which are able to carry out several metal-cutting operations in succession) can result in bottlenecks elsewhere in the plant, so these machines often end up working well below capacity (Gallacher and Knight, 1973).

During the 1960s and 1970s, however, a different type of plant layout was championed for batch production. This was known as 'Group Technology' (GT).[2] GT involved grouping machinery on the basis of what machines were necessary for the production of a range of broadly similar products or components within·a component 'family'. The result is the *group layout* featured in Figure 6.1(b). Such a configuration permits the production of different products requiring different sequences of machining. By creating small production cells, each able to produce a more limited range of products than can be produced among a collection of machine shops, economies can be made in the use and turnover of fixed capital.

There are two major sources of economy here which are said to result from the use of GT. They are those derived from (i) simplified flow-paths for materials and (ii) reduced times for setting-up

(i) Complicated

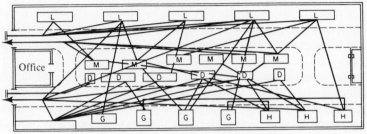

N.B. 'Functional layout' – machines grouped by process type

(ii) Simplified

(By laying out in family machine 'groups')

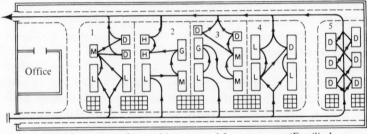

N.B. 'Group layout' – machines grouped for component 'Families'

FIGURE 6.2   *Simplification of material flow system ('functional layout' vs 'group layout')*

SOURCE   *Burbidge (1975)*

machines. By grouping machines on the basis of component families, the flow of work becomes more regular, whilst retaining flexibility in the sequence of operations within the particular group. Also, the increased proximity of different machines required for the production of a particular component reduces the distance over which the work has to be transferred between operations (see Figure 6.2).

Using this form of layout, however, entails the grouping of products into families on the basis of a similar set of machining operations. This is a prerequisite for planning the composition of machines which make up each group. However, the classification of product types also has a further importance. By reducing the variety of products which are worked on a particular machine the setting-up time can be reduced because of the narrow differences between components of the same family. Also by planning the sequence of components being machined, as discussed later, changeover times for resetting machines between components can often be reduced even more.

So far it might be possible to interpret cell grouping and family machining as a half-way house between functional layout and the line organisation of Fordism. That is, it is compatible with a tendential movement towards Fordism given the limitations imposed by product variety. However, there are several respects in which GT differs significantly from Fordism.

The first of these is associated with changes in the size of batches. In functional layout, production is organised around the concept of the minimum economic batch quantity (EBQ). This represents the supposed optimal quantity in terms of the unit cost of a batch of identical components, balancing preparation costs (ordering and setting up) and carrying costs (storage expenses, along with the fixed capital tied up in component stocks). Hence the volume of stock held is directly related to the size of the EBQ. As this stock levels falls low, a new batch of parts is produced in the 'economic' volume.

Where Fordism would suggest a rationalisation of products in order to achieve lower variety and longer runs in production, GT proposes a *reduction* in batch sizes. In the case of functional layout using stock control and EBQs there are multiple cycles of production for different components. This is because the different components are scheduled for production at random times depending on the stock level of a particular item falling below a minimum number. In contrast, GT reduces the average size of batches by replacing the multiple cycles of stock control by a single cycle of small batches of components made to a regular production schedule (Burbidge, 1974, p. 78). In this way, it is argued, a reduction in stock levels can be achieved. This is augmented by shorter waiting and handling times which become possible when adjacent cells make the different components for a product in parallel. This has the advantage of increasing the flexibility of batch production thus reducing the time taken to produce a particular order, or batch of products. Similarly the form of production control associated with GT contains the potential for *a reduction in the lead-time* for producing new or changed products and a *reduction in component obsolescence* occurring on changeover since components are produced in product sets. Hence, there is a second aspect of the neo-Fordist character of GT: its appropriateness to variety and change in products.

Crucially, however, the feasibility of this form of production organisation depends on a reduction of the time in setting-up machines without which increasing setting-up costs would severely reduce the economic benefits of reduced stock levels and shortened lead-times. Originally, this was carried out by using such techniques as: attention to the sequence of products (minimising the amount of setting-up changes between products); creating 'families' within tools; pre-setting of tools in tool-holders; and later the use of co-ordinate setting techniques and digital component-size monitoring to avoid using stops. Of course, the use of numerically controlled machinery to reduce setting-up time has become more important recently as they have diffused more widely into this sector. Yet, the forecast coming together of numerical control and GT has not occurred to any great extent (Swords-Isherwood and Senker, 1978, p. 2).

Finally, both line- and function-layout can be accompanied by increased specialisation in the division of labour. By allocating particular tasks to workers on work-stations or within particular shops, skills can be restricted or eliminated. Within machine shops, worker-resistance in the form of job controls has had a definite protective effect on the erosion of skill.[3] Yet job controls are limited in most cases to particular skilled 'trades'. The work of other semi-skilled workers in shops and workers in mass-production lines is much more specialised and, therefore, open to routinisation and degradation.

The 'cell' type of organisation favoured by GT, however, can cut across the forms of work organisation and job control in machine shops which are characteristic of occupational specialisation based on skilled trades. This is a result of workers being assigned to cells capable of several types of machining; the work-roles here, therefore, demand the capability to carry out a range of operations, some of which will require skills different from others. The degree of variety in the tasks associated with cell production served as the platform for one proponent of GT to link its use with job enrichment and socio-technics (Edwards, 1974a, 1974b).

Despite being very fashionable as a technique for reorganising production in the late 1960s and early 1970s, Group Technology now is very much out of favour as a complete package. A survey of machine shops in Britain showed that while NC has become popular, interest in GT has declined with some of those who showed some early enthusiasm for GT returning to a more conventional machine-shop layout. (Swords-Isherwood and Senker, 1978, p. 45). It is clear from the case studies presented by this survey that GT solved some problems (such as reducing throughput times) more than others (such as low machine utilisation and the costs of training skilled workers in several machining skills). Given the managerial complexities of planning the flow, grouping and sequence of products, as well as the loading of products between different cells, it is hardly surprising that the potentialities of GT proved hard to achieve in practice. However, with the increasing sophistication and reliability of NC it became possible to achieve some of the benefits that GT promised through reduced setting-up times. This did not remove the organisational problems of machine shops, which remained to be faced again later. It entailed a shift of focus from the organisational level to that of the individual machine. Hence the growth of multifunction machining centres where workpiece transfer could be eliminated. Put another way, the complex of problems involved in the transformation of small-batch manufacturing became seemingly more approachable by concentrating on the technological dimension. The wholesale reorganisation of machine shops demanded by GT along with the consequent costs of worker resistance, upheaval and likely price of failure, became avoidable much to the disgruntlement of GT proponents who attacked this as the result of a 'failure in British management' (Burbidge, 1978; Edwards, 1980).

In the following section, then, we will describe changes in the technological dimension of small-batch engineering with flexibilities in machine-control systems constituting the important element. These technological developments, notably that of numerical control, occurred in parallel with Group Technology, although their widespread diffusion is a more recent and continuing phenomenon. It is only later, when attempts are made to link numerically controlled machinery into automated manufacturing systems that the full impact of these organisational problems recur. This time, however, the planning problems can be eased to a greater extent by the availability of computer-aided manufacturing systems (CAM). Also, the rationalisation of product design brought about by the use of computer-aided design in the drawing office to customise standard product designs can aid the grouping of similar components for cell-based production in 'Flexible Manufacturing Systems'.

TERTIARY MECHANISATION OF MACHINE TOOLS

In earlier chapters by tracing the development of Fordism, we noted particular characteristics in the development of Fordist mechanical control systems. By using mechanical fixtures (to hold the workpiece in the correct position for machining), jigs (to guide the cutting tools) and stops (to limit the depth and dimensions of cut) this tooling permitted certain work previously carried out by skilled workers to be performed by semi-skilled workers, but only after the machinery had been initially set up by the former. Furthermore, the factors of skill shortages and the need for increasing precision have been stressed in this account as significant elements of the 'deskilling' dynamic besides economic considerations of labour costs. However, the rudimentary nature of such control mechanisms has restricted their use, in the main, to simple, repetitive operations. In the case of transfer lines, where such operations are carried out in succession, mechanised transfer devices have been used to move parts between the 'dedicated' machines.

Yet for more complex work, involving a variety of products, the mechanisation of control systems on machine tools has taken a different path. Noble (1979) describes the history of such technologies. In the 1930s and 1940s the use of tracer mechanisms involved the use of patterns or templates to guide the cutting tool, thereby reproducing the contour of the template in the workpiece. Here the self-acting nature of the machinery is not restricted to the production of identical, or very similar, items as is the case at Fordist control mechanisation. Changing the template permitted a different cut to be made, hence a significant degree of versatility was retained alongside machine-tool 'automation'. However, the mechanical control mechanism was an early form of machine control. The drawbacks of producing templates and needing several of these for the different surfaces of the workpiece were eventually overcome by using electrical control mechanisms. The development of new electrical sensing and measuring devices, as well as precision servometers controlling mechanical motion (a development of gunfire control technology developed during World War Two) served as the technical means for this.

Noble's case-study examined one particular choice in the process of developing electrical control of machine tools. This was between an *analogue* form of programming, in which the machine recorded the initial machine motions made by the operator, and a *digital* form in which the sequence and details of operations were specified numerically and stored usually on a punched paper tape. By replaying the programme, identical parts could be produced. In the 'record-playback' technology, programming was commensurate with the skills of existing machinists. With NC, however, these skilled operations need to be converted into a digital programming code – 'parts programming', as this procedure is called – and this dispenses with the manual controls of machining and so makes one aspect of the craft-skills redundant, namely what we have previously referred to as hand-eye co-ordination.

As was noted earlier, Noble wishes to argue that the choice between these control technologies reflects the social relations of production. The 'horizontal' relations of production (for instance, between different firms) favour the development and use of NC over record-playback because the complexity and expense of the former favoured larger military contractors over smaller machining shops. The 'vertical' relations of production (that is, between capital and labour) were mirrored in the use of NC since this technology permitted deskilling. In this way Noble elucidates the choices in technological development and how the social relations of production of capitalism constantly precludes the development of other possible lines of technological change.

It is not clear that the record-playback system did indeed perform as well as the NC system. As far as military requirements are concerned, NC seemed to them more appropriate to their requirements, as Noble admits. Thus the complexity of aircraft parts, the possibilities of standardisation of programs (having strategic and cost-cutting value) and the potential which NC software held for rapid mobilisation and design change all mitigate against record-playback. What is more, the development of NC machinery does not inevitably lead to a distinct division of labour between programming and operating (Sorge *et al.*, 1981; Jones, 1982a). Nor did the adoption of NC machinery preclude the later construction of NC machinery having record-playback characteristics (Marsh, 1981a, p. 545; Boon, 1981).

What numerically controlled (and for that matter record-playback) machine tools do represent, however, is a different form of 'automation' to the transfer line. The defining characteristics of this form of control lies in the ease of reprogramming the control system with different instructions to enable the machine to carry out different operations. What these technologies effectively achieve is a drastic reduction in the setting-up time of machine tools (which was of crucial importance in the implementation of GT). This reduction is a result of: (i), the ability to program (now comprising a major part of setting-up activity) *off* the machine – thereby reducing the 'down-time' when a machine is not utilisable for production; (ii), the ability to store a suite of tapes or programs from which a number of repeat items, or variations of these, may be easily made as required.

The mechanisation of machine tools can be characterised in Bell's scheme both in terms of the sophistication of control mechanisms and the mechanisation of handling (transfer and loading) functions as follows: the Figures 3.1 and 3.2 represent developments in the mechanisation of machine tools under two headings: the early 1960s and the 1970s. The major difference between the two lies in the sophistication of the control systems. In Figure 3.1, increased sophistication of this function in the period before the 1960s was generally accompanied by an increased inflexibility in operations and sequencing. After that time (Figure 3.2) newer systems of control were used which did not rely on the manual setting of dials, switches, stops and cams. Although some manual inputs are still necessary, different categories of electrical control system (numbered 5 to 8) raise further the level of control mechanisation. In this period, because setting-up began to take the form of programming – using plugboards, paper and magnetic tapes, or direct inputs from a computer ('Computer Numerical Control' – CNC), it became pro-gressively easier to change the details of operations between different products.

The two phases in the development of control mechanisation in metal-cutting machinery identified by Bell correspond closely to our distinction between Fordism and neo-Fordism. In the early phase, increased sophistication of control mechanisation, being accompa-nied by inflexibilities in machine operation, had to be associated with product standardisation and volume production. The latter phase is characterised by the progressive uncoupling of the level of mecha-nisation and the scale of production.

Given this distinction, the effects of higher mechanisation in mechanical engineering is variable. Under Fordist mechanisation the relation between standardisation and the level of *control* mechanisa-tion has the effect of increasing the proportion of less skilled tasks. This is despite the contrary trend of increasing need for skilled tooling, setting-up and maintenance tasks, whilst in the case of the mechanisation of *transfer* (loading and handling) functions, it is not obvious what specific effect this will have since it is dependent on the actual division of labour as to whether skilled or semi-skilled labour is displaced.[4]

Ultimately Bell noted that as a general rule the mechanisation of *control* displaces skilled tasks, whereas the mechanisation of *transfer* functions displaces unskilled tasks. Under Fordism, because of the link between scale and level of mechanisation, the overall effect is one of reduction in the proportion of skilled tasks and an increase in unskilled ones. In the newer, neo-Fordist phase of mechanisation, the potential effects of technical change on skill and work organisation are much more difficult to identify. We have seen that this difficulty has led some writers to adopt a contingent model of the relation between technical change in machine tools and skill (Jones, 1982a) or work organisation (Wilkinson, 1981). The eventual outcome of technological change is seen very much as the outcome of a process of negotiation surrounding the introduction of the technology.

Examining Bell's model of technical change in the 1970s (Figure 3.2) he identifies four major levels of control mechanisation. Unlike Fordist mechanisation, successive developments in this dimension are accompanied by an increased *versatility* of the machine tool's operations, resulting from the programmability of the control system. The limited nature of the memory, reducing the number of possible programs which are available from machines in category 5 is overcome in category 6. In this case the use of paper or magnetic tape allows a virtually unlimited memory, which is accessible by manually loading the requisite tape on to the machine. In category 7, the programs are stored in a computer memory which permits a more rapid access, and thus program change, although in all the cases the machine itself will still usually require changes in tooling between different programs. Further limitations to the versatility of the control system being realised result from the limited mechanical capabilities of the machine itself. For example, some machines are more suitable for particular types of operation than others by virtue of the range of tools or size of workpiece which can be accommodated. To an extent this has been overcome by the use of machining centres which have a wider range of tools and capabilities than ordinary lathes or milling machinery. Machining centres also have the advantage of overcoming the need to transfer workpieces between different machines since several operations can be performed on the one machine.

Bell also described a variant of direct computer control (Category 8) incorporating rudimentary sensory feedback capabilities. 'Adaptive' control systems would monitor parameters such as workpiece dimensions, cutting speed, etc. and compensate for effects like variations in tool sharpness or the flexing of certain alloys during machining.

When Bell carried out his study there were only a few examples of machines in categories 7 and 8 and most of these were still under evaluation. The majority of machines in use were those in categories 5 and 6. Hence he saw the general effect of NC machines as reducing the need for skilled workers as operators, by means of the separation of programming and operating. By extrapolation, the Direct Numerical Control (DNC) machines of categories 7 and 8 would amplify that effect, given that DNC, in most cases, consisted of a large central computer storing the NC programmes which were then routed electronically to an array of NC-input machines. The counter-effect to this though was in the mechanisation of handling, which would tend to displace less-skilled workers if they were previously carrying out these transfer functions. Where the latter functions were carried out by skilled workers (for example, loading and unloading their machines) then the usual effect would be to increase the skilled content of their work by removing unskilled handling functions. Ultimately then, Bell identified the likely direction of development in small-batch mechanisation as a central computer controlling NC-programmed machine-tools linked by transfer devices which allow flexible routeing of workpieces between the different machines. The net result of increasing mechanisation therefore, would be, first, a shift to using semi-skilled machinists on NC machines, followed by a reduction of the latter with the increased mechanisation of transfer culminating in an integrated manufacturing system, with few operators but an increased emphasis on planning and maintenance functions (Bell, 1972, p. 77).

In this respect Bell's scheme follows the predilection of many writers on technological change by extrapolating the labour-displacing effects of mechanisation to a labourless fully-mechanised production system. However as we have been at pains to point out, such a one-sided and determinist view of technological change hides the contradictory dynamics of the production of profitable commodities. It severely compresses the process of diffusion of technology during the short term, where severe over-capacity makes investment in such new technologies highly risky. In the longer term there are different options for the development of control mechanisation other than one of the centralised control of highly integrated system as has been the goal at high levels of Fordist mechanisation such as in oil and chemical plants.

As we have pointed out, there is no necessity for mechanisation to be as advanced along all dimensions of transformation, transfer and control. Since Bell's work, decreases in the size and cost of computers in the last decade have led to the construction of NC machinery with integral computer control – 'Computer Numerical Controlled' (CNC) machines. Such machinery does not have to incorporate a form of mechanised transfer facility in order to represent a substantial development in machining capability. Indeed for many operations the mechanisation of inter-machine transfer would be difficult. Whereas CNC can be used, like NC, with separate programmers and operators, CNC also allows the reintegration of operating and programming, greatly facilitating both editing and programming on the machine. CNC tools also go with the development of powerful software which can reduce the time taken to construct and 'prove' programs as well as to make modifications to existing programs. Computer Numerical Control machinery, then, has a capacity for *decentralised* usage which is very different from assumptions of central computer control characteristic of DNC systems. Now *both* modes of CNC usage can be found in practice, and CNC machines can themselves be linked to a central computer. So what factors influence the mode in which CNC is used?

In a comparative study of the diversity of forms of work organisation associated with the use of CNC technology, Sorge *et al.* (1981) try to disentangle what they refer to as the different logics of CNC application. They note that these different logics are analytically distinct but that they sometimes interact or conflict with each other. They are associated with these factors: company or plant size; batch size or time; type of cutting or machining; national institutions and habits of technical work, management and training; socio-economic conditions (for example, raw materials shortages, limitations to market size and slow growth).

As plant and batch size increase they note a correlated increase in bureaucratic structure (that is, a separate planning department) and increased polarisation of skills (between operators and setters/ programmers). The phenomenon is modified by the time it takes to write a program. In general, the greater this is, the more programming becomes differentiated from operating. They found that such task differentiation had more to do with management's attempts to achieve maximum utilisation of expensive machinery than the control benefits of using a separate programmer. Note also that the separation of planning and operating did not necessarily imply a polarisation of skills, because, where programming was complex and therefore took a lot of time, the machinery itself was often sophisticated (for example, multi-function machining centres) and required skilled operation. On simpler work where polarisation is more likely under NC, the sope of CNC for merging programming and operating functions becomes greater (Sorge *et al.*, 1981, p. 6).

National differences also were important. They noted a consistently greater use of shop-floor- and operator-programming in West Germany as compared with Britain. This is reflected in the selection of equipment and control system options which are appropriate to shop-floor- and operator-programming, as well as in significant national differences in training, labour levels and work organisation policies. For example, they found that in Germany, technician training took place *after* craftwork apprenticeships rather than separately. In Britain programming was more likely to be a white-collar occupation wheres in Germany rotation between machine-based and planning department programming occurred (Sorge *et al.*, 1981, p. 8). This observation is in line with the findings of both Jones and Wilkinson that management's attempts to separate programming and operating functions were a constant problem in Britain, being actually *detrimental* for such things as quality and scrap rates.

By using CNC rather than NC machinery operator programming was made more feasible. This 'technological capability' of itself cannot account for the eventual organisation of work with CNC machinery, since CNC is also compatible with a separate programming and planning department, as is common with NC. It might be attractive, given the manifest diversity in forms of work organisation associated with CNC, to adopt a strong 'shop-floor politics' position arguing like Wilkinson that the eventual polarisation in skill and programming functions will depend on contestation between different parties. To do this, however, would concede that the ultimate determination of the structure of the labour process occurs at such a level. Similarly, Jones's analysis might lead us to expect a diversity of organisational forms as a result of the mutual interaction of product and labour markets, technology and labour organisation. However, even this broader spectrum of variables does not give any feeling for dominant directions in technological and organisational change. There seem to be no general trajectories of development, like that of Fordism, along which new technologies and forms of work organisation are being directed. Yet if we take Sorge *et al.*'s fifth factor – the *socio-economic conditions* of these changes, it becomes possible to relate these changes in the labour process to the broader political economy.

In their analysis, Sorge *et al.* note the specific conditions of competition in mature Fordism which become associated with many large producers competing in large markets. Firms have increasingly had to cater for particular market niches rather than monopolising the large or mass markets available in the past phase of Fordist growth. Market expansion or increase in market share requires increased variety and technical change in products. More individualised and customised commodites increase the complexity and range of parts in production.

This acceleration in product complexity could be most adequately accomplished by designing more complex parts and components. Yet the requirement would then exist for a much greater number and complexity of operations to be carried out to make each component. They point out that it is for this reason that the diffusion of CNC machinery has a *qualitative* importance over and above increased productiveness defined in a quantitative sense. Thus the relationship between technological change in machinery and change in products is an organic one which is difficult to express quantitatively in the form of economic cost per unit of output. The complexity and precision of metal-cutting made possible by CNC machines would have been difficult, if not impossible, to produce on conventional machinery without using highly specialised and costly tools (Sorge *et al.*, 1981, p. 11).

This is not to under-estimate the significance of cost reduction with NC and CNC. In the previous section we noted that the economic importance of reduced setting-up time (including programming time) between different production runs along with the further economies attainable with smaller batch sizes was a result of reduced levels of stock and work-in-progress. We have argued that some effects similar to those claimed for Group Technology might also be claimed for CNC machinery. One recent example here which has attempted to forecast the potential effect of 'flexible' technologies on the economics of batch production is the work of Ayres and Miller (1981).

Ayres and Miller analysed the returns of the 1977 US Census of Production for selected metal-working industries which are representative of the range of metal-forming processes (Table 6.1). Comparing these industries, in terms of value added (labour and capital goods cost) per unit of raw materials gives the unit processing cost per dollar of materials used for the different industries. The lowest cost ratio in the group (0.6) for metal cans, an example of mass production, is less than one quarter that of 'special dies and tools, jigs and fixtures'. That is, the small-batch industry spends over four times as much on capital and labour as the large-batch metal can industry per unit of output.

If the difference in cost of materials processed in the various industries is taken into account, with some industries using high quality tool steel or expensive alloys while others only use ordinary carbon steel, they estimate that the unit costs are about ten times higher for custom production in a job-shop environment than they

TABLE 6.1   *Unit Cost Index of selected steel products, USA, 1977*

Product	1977 Value of shipments $ \times 10^6$	1977 Value added $ \times 10^6$	Cost ratio
Special dies and tools, jigs and fixtures	3909.6	2791.5	2.497
Cutlery	711.1	494.0	2.276
Carburettors, pistons, rings and valves*	1362.4	900.5	1.948
Hand and edge tools	2220.4	1403.3	1.717
Screw machine products	1759.0	1015.8	1.367
Ball and roller bearings	2575.3	1478.5	1.347
Bolts, nuts, washers and rivets	3311.0	1842.0	1.254
Automotive stampings	9708.7	4642.0	.916
Iron and steel forgings	2806.8	1300.3	.863
Metal cans	8097.7	3044.6	.600

* Not primarily for auto engines

$$\text{Cost ratio} = \frac{\text{Value added}}{\text{Purchased materials}}$$

Purchased materials approximated by 'value of shipments minus value added'
SOURCE   *Ayres and Miller (1981).*

SOURCE   *Ayres and Miller (1981).*

are for mass-production operations. (This figure is in line with the estimation made by Cook, 1975, and quoted earlier). Now, on the basis of assuming a two-thirds reduction (derived from a survey of the potential cost reductions resulting from technologies now in experimental use) in the differential between unit costs of mass and batch production, they are able to compute the potential cost-savings of small-batch mechanisation. Taking into account the variation in the value of materials processed they estimate the final price reductions as in Table 6.2. The two-thirds reduction in processing costs reduces overall costs by anything from 7 to 30 per cent, being highest for those products most likely on the whole to be produced in smaller batches.

The potential effect of this phase of neo-Fordist mechanisation, then, is a substantial reduction in the cost of products within the batch metal-working sectors. Along with the reduction in manufacturing costs, there are also qualitative changes in the nature of products implying a significant role for new 'flexible' technologies in the transformation of small-batch engineering. The realisation of such a potential is a different matter. In order that the effective use of such computer-controlled machinery should be achieved, far-reaching changes in the organisation and administration of machine-shops are necessary. What is currently described as 'Computer Aided Production' (CAP) would have to cover technological, organisational *and* informational transformations in the present structure of machine shops. We can see examples of this process being attempted

Table 6.2 *Potential cost reductions in selected steel products*

Product	Potential savings* ($ × 10^6)	Effective price reduction* as %
Special dies, tools, etc.	1172	30.0
Cutlery	204	29.0
Carburettors, pistons, etc.	312	23.0
Screw machine products	284	16.0
Ball- and roller-bearings	409	16.0
Bolts, nuts, etc.	480	14.5
Automobile stampings	851	8.8
Iron and steel forgings	197	7.0
Metal cans	0	0

* Compared to metal cans

SOURCE *Ayres and Miller (1981).*

in leading firms at present. In this last section we investigate some of the dynamics of the transformation.

COMPUTER-AIDED PRODUCTION

Whilst most of the labour process studies of NC machinery have looked at the level of individual free-standing machines and the division of labour between programming and operating, there have been moves to create integrated metal-working systems. The integration of several separate machines by *flexible* mechanised transfer systems, using computer-controlled pallets or occasionally robots carrying individual workpieces, has led to the experimentation with *Flexible Manufacturing Systems* (FMS). Information technologies are being used to plan and co-ordinate flows of materials, stock levels and production schedules, in what computer sellers call computer-aided manufacturing (CAM). At the design and drafting stage, where the specification of each product and its components takes place, information technologies have in use from the late 1960s. Such computer-aided design (CAD) systems, and CAM, represent the extension of mechanisation to conceptual work. Ambitious attempts to link all these systems together to form Computer-Aided Production (CAP), 'Automated Small-batch Production' (ASP) 'Computer Aided Engineering' or 'CAD/CAM' systems are also mooted. All these systems are in the process of development and so inevitably there must be some uncertainty as to their final configurations. Nevertheless, recent studies have drawn interesting conclusions as to the likely direction in which such systems will progress.

Flexible Manufacturing Systems (FMS)

In its simplest form, a flexible manufacturing system consists of several NC or CNC machines linked by a flexible transfer mechanism which is capable of varying the sequence of workpiece-transfer between the different machines. This would be called a 'flexible machining cell' and might resemble Figure 6.3. In this case a robot is used to perform the loading and transfer operations. More complex Flexible Manufacturing Systems might consist of a larger number of computerised metal-cutting machines with robot-loading and a flexible pallet transfer system as in Figure 6.4.

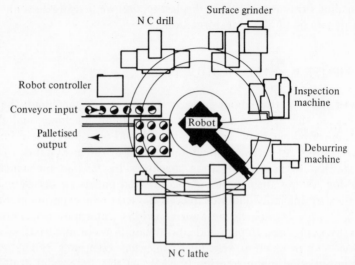

FIGURE 6.3 *Flexible machining cell (multi-product family random batch production)*

SOURCE *Technology Week, no. 2, 13 February 1982.*

In principle, the design of an FMS is relatively straightforward. Versatile, computer-controlled machine tools perform the required operations on different workpieces which are transferred between the machine tools by a transfer and loading system under central computer control. Although there are several examples of small-scale FMSs in operation, the attainment of advanced, unattended FMSs has been a goal of state-financed R & D in Britain (via the Automated Small-batch Production (ASP) programme of the Department of Industry), Japan (in the Unmanned Metal-working Factory (MUM) project under the control of the Ministry of International Trade and Industry) and in the USA the Air Force is funding a program called ICAM (Integrated Computer-Aided Manufacture).

The model used here, or 'technological dream' as Noble might describe it, is that of an integrated engineering plant so highly mechanised as to resemble a continuous process plant. As one British government-sponsored committee enthused:

135

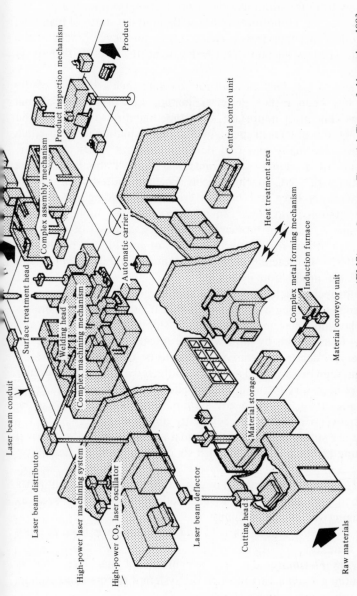

Laser beam conduit

Laser beam distributor

Surface treatment head

High-power laser machining system

High-power CO₂ laser oscillator

Raw materials

Welding head

Complex machining mechanism

Automatic carrier

Laser beam deflector

Cutting head

Material storage

Material conveyor unit

Induction furnace

Complex metal forming mechanism

Heat treatment area

Central control unit

Complex assembly mechanism

Product inspection mechanism

Product

FIGURE 6.4 *A Japanese flexible manufacturing system (FMS)*   SOURCE *Financial Times, 3 November 1983*

(NOTE Another version of this diagram appeared in *New Scientist*, 30 October 1980. Then the diagram included five human figures, two in the Central Control Unit, and one each in the Heat Treatment Area, the Machining Mechanism and the Assembly Mechanism. The *Financial Times* version, with all the figures removed, compounds the illusion of the 'completely automatic' factory.)

Advanced, fully automatic (ASP) systems completely remove workers from the arduous side of production (tempo, noise, heat, fumes, danger, monotony) and raise them to the status of planners, controllers or supervisors – analogous to the development of the chemical processing industry. (ASP Committee Report, 1978, p. 4)

Needless to say the timescale for realising such a 'factory of the future' has been rather under-estimated. Leaving organisational problems and, most important, economic viability to one side, the technological problems of such a major transformation in small-batch production are many and varied. These are not restricted to the integration of machinery and transfer systems. Limitations in the automaticity of CNC machine tools which favoured their use by skilled workers have not been overcome by adding mechanised transfer systems: problems such as tool wear, tool breakage, on-machine inspection, lubrication, coolants, disposal of swarf are stubborn problems in the mechanisation of small-batch production (Sims, 1982). In practice then, skilled human intervention is still required for production in a FMS.

A major technological problem is that of reliability. The drawback of a completely integrated system is that if one component breaks down, the whole system may have to be stopped. Therefore, as well as trying to avoid this by improving the reliability of the constituent machinery, a degree of *redundancy* has to be built into the system. Work can then be re-routed around the affected machinery. A similar feature is necessary for maintenance and training purposes (Gerwin and Leung, 1980, p. 7). In any system, then, a balance has to be struck between redundancy, reliability and cost.

There are still limitations to the actual flexibility of FMSs. Gerwin and Leung (1980, p. 15) distinguish six different types of flexibility:

- *Mix Flexibility:* processing at any one time a mix of different parts which are loosely related to each other in some way such as belonging to the same family.
- *Parts Flexibility:* adding parts to and removing parts from the mix over time.
- *Routeing Flexibility:* dynamic assignment of machines; that is re-routeing a given part through the system if the machine used in its manufacture is incapacitated.
- *Design Change Flexibility:* fast implementation of engineering design changes for a particular part.

- *Volume Flexibility:* handling shifts in volume for a given part.
- *Customising Flexibility:* processing different mixes of parts on different FMSs in the same company.

Making FMSs flexible along all these dimensions would of course be expensive. Although only certain degrees of flexibility may be required at any one time, they found that the effect of market changes was to change the crucial aspects of flexibility demanded from the FMS. Whereas they argue for FMSs to be designed with 'flexibility responsiveness' in mind, it is clear that the more highly mechanised and integrated the FMS the less flexible it becomes.

When consideration of reliability, quality monitoring and the need for sophisticated maintenance and trouble-shooting skills are also taken into account, the move towards large 'Automated Small-batch Production' systems becomes more problematic. We now find much more modest proposals put forward in, for example, the British ASP programme (Hollingum, 1980; *Production Engineer*, 1981; Sims, 1982). The accent is on smaller, more simple systems.

A recent survey (*Computing*, 14 October 1982) of the thirty-three flexible systems in the United States revealed that none were working to the operator's complete satisfaction, a few apparently being downright 'disaster areas'. Similarly a German study (Blumberg and Gerwin, 1981) argues that highly complex integrated systems such as these were as yet beyond the *organisational* capacity of most firms. Rather than the ideal of an 'automated factory', often depicted as a 'smoothly operating system free of human variability . . ., it may well be a nightmare of uncontrollable problems requiring unanticipated human intervention at critical points and unexpected upheavals in the workflow structure' (Blumberg and Gerwin, 1981, p. 23). Increasing complexity and integration of an FMS system makes it difficult to monitor defects in workpieces between operations. By the time a defect is discovered, other parts being worked or the machine tools themselves can become damaged. Similar findings are reported by Shaiken (1980, p. 7) in a study of the pioneering system at Caterpillar Tractor, Illinois. He describes how it took four years from the date of starting construction for even marginal operation to be achieved. In the first few years of operation, problems resulted in a utilisation rate of only 20–40 per cent. Yet now the engineer still complains of 'nagging problems' in the system which were visible from the number of machines broken down during a recent visit!

In order to solve problems such as these, Blumberg and Gerwin point out that, as the complexity of the system increases it is often necessary for firms also to increase the amount of human intervention – thus partially reversing the trend towards 'automation'. They conclude that the majority of companies are likely to gain more success by using simpler technologies which may be gradually built up in complexity whilst still retaining human labour as an essential part of the production process. Boon (1981) comments that the rigidities and inefficiencies which result from increasing complexity of FMSs make them more appropriate for large firms having a highly rationalised product range. Smaller firms with more diverse products attain better economies in practice using skilled workers operating smaller FMSs or machining cells.

CAD, CAM and CAP

If FMSs are characterised by the attempt to realise a *technological* solution to machine utilisation and 'throughput times' in small-batch engineering. Computer Aided Manufacturing (CAM) is an attempt to effect an *organisational* and *informational* transformation. In small-batch production, we have suggested, all three of these aspects are interrelated. Solutions which focus on only some of these become beset by problems from the neglected aspects.

If we examine the range of tasks and operations involved in running a machine shop (Figure 6.5), it becomes clear that the actual tasks of machining and transfer of materials are heavily dependent on prior organisational and informational tasks of work analysis and planning, production control, tool and workpiece availability. Similarly, the design of products and components including the production of drawings for use in machine programming and machining, ultimately determine the need for materials, tools and machinery. Changes in small-batch engineering are beginning to occur at these conceptual levels.

We have defined neo-Fordism as a change in the organisational element of the labour process from an *hierarchical* to an *informational* one. In Fordist production systems a bureaucratic and hierarchical supervisory structure is used to plan, co-ordinate and control production organised around the interrelated tasks of line-workers. Yet neo-Fordist modes of organisation involve some decentralisation of these functions. This form of organisation is only feasible within an overriding plan of production. In this sense, the informational dimension of production becomes central.

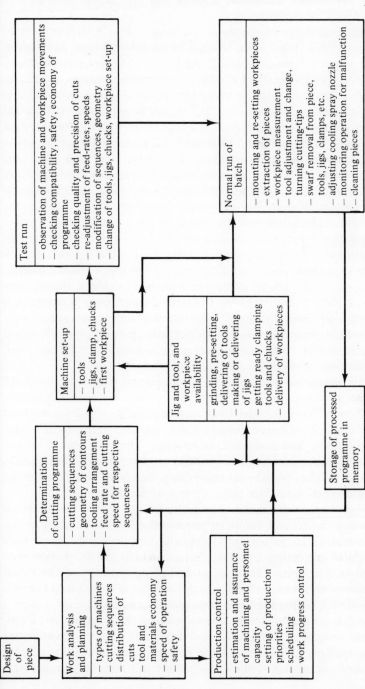

FIGURE 6.5 *Conception and execution of machining tasks*

SOURCE: *Adapted from Sorge et al., 1981: p. 55.*

The availability of computerised information or data-processing has permitted step-by-step changes in the informational activities of manufacturing firms. Nygaard and Fjalestad (1981) describe three stages in the development of corporate computer systems. In the *first stage* a large mainframe computer is installed in a 'Computer Department' and data from other departments of the organisation are collected and converted into the appropriate format to be processed in the central department. In the *second stage* the other departments are reorganised to fit around the data-processing requirements of the central computer. All departments now produce standardised data for the computer to process and act upon its standardised output. But in the *third stage* the walls between the Computer Department and the rest of the organisation are penetrated by communication equipment – linking the Department's equipment to local computers, electronic typewriters, video screens (with or without associated keyboards), analogue-to-digital converters, etc., throughout the organisation. As they put it, 'the organisation is infiltrated by computing equipment. It becomes a network of human beings, machines and information-processing equipment, linked together by an increasing proportion of electronic communication links and a decreasing proportion of human communication links' (Nygaard and Fjalestad, 1981, p. 224).

This process has been taking place for several years, and is thus not a new phenomenon associated with recent advances in micro-electronics. Since the mid-1960s, major firms have used centralised computerised data-processing (Stage 1) for such functions as wages, sales ledgers and inventory monitoring. These organisations are currently moving towards the establishment of common data bases as in Stage 2. The common data bases may then be accessed via remote 'unintelligent' terminals for a variety of functions.

Where micro-electronics has had a major effect has been on reducing the cost of computer *hardware*. This has resulted in the development of 'intelligent' terminals, terminals with an in-built computer-processing capability. Such terminals may be dedicated to a particular use, such as planning production or material-requirements, but may be able to draw on the common data base for information as required. In some cases a central processor with a powerful computer can be shared between these terminals ('distributed processing'). Nygaard and Fjalestad refer to these latter developments as the third stage in the structural development of corporate computer systems. Essentially this third stage represents the beginning of a thorough transformation of the *conceptual* functions of production.

In the case of design and drafting, the conventional routine consists of a designer or drafting worker drawing a product or component and the resulting drawing, with details such as sizes, materials finish, etc. will be distributed to machinists, NC parts programmers, project engineers and methods engineers. They, in turn, have to interpret these drawings, extracting the relevant information for their own purposes. The resulting information will then be translated into production schedules, NC programs, raw material requirements and tooling needs, which are used to organise various elements of production.

The first part of this process, designing and drafting, has been the site of increasing use of computer-based graphics systems. Initially, these were limited to large mainframe computers (as used in Stage 1), and thus were expensive systems to buy and use, which restricted them to high value sectors such as aerospace and electronic circuit design, and mass-production sectors like motor vehicles. Mini-computers and micro-computers, by lowering the cost of hardware, along with cheaper storage display tubes and powerful software permitted a reduction in cost of Computer Aided Design systems. It became economically feasible for mini- and micro-computer-based CAD systems to be used in other sectors of engineering.

Whereas CAD systems have become more sophisticated in recent years, having capabilities such as rudimentary three-dimensional and solids modelling and incorporating statistical procedures to analyse such things as stress factors, these do not account for the most widespread use of CAD. CAD might be better thought of as computer-aided-*drafting,* since it is in this mode that most existing systems are used (Arnold and Senker, 1982). This is especially the case for small-batch engineering, given its lag behind other sectors of engineering in the use of CAD.

CAD is particularly appropriate in small-batch engineering for the modification and manipulation of existing designs, more than for designing entirely new products. In this resepct the use of CAD facilitates the customising of products – a feature that characterises much of the work of the mechanical engineering industry – as well as reducing lead-times for new orders. The diffusion of CAD seems to have been accelerated by the shortage of drawing-office labour, which was common until very recently. In the recent past, and more so in any future period of expansion. CAD will be a technological substitute for the lack of skilled workers rather than an attack on existing ones. What is more likely, according to Arnold and Senker, is that drawing-office workers *without* a knowledge of engineering gained on the shop-floor will be most at risk from CAD; this in turn threatens the training route, and therefore the supply of more experienced CAD operators in the future (Arnold and Senker, 1982, p. 35).

Where firms have attempted economic calculations to justify the introduction of CAD, this is often done on the basis of labour displacement. It is not clear, though, that labour costs are the most relevant consideration. Factors which influenced many managers in their decision to press for a CAD facility were related to competition. Better presentation of tenders, shorter lead-times, the need for increasingly complex products, the ease of meeting customers' product requirements, increased accuracy and clarity of drawings all proved to be important but unquantifiable aspects of the diffusion of CAD. As we have already said, this is a crucial aspect of the forms of competition under neo-Fordism: heterogeneity in product markets and technological competition (in both products and production processes) replace the large homogeneous mass markets and relatively stable products of the oligopolistic Fordist phase of accumulation. The cost-saving justification of labour displacement may appeal to accountants during a depression. Yet the ability to expand in an upturn and survive competitively against other capitals demands counter-cyclical investment and the implementation of, and experience with, CAD systems *before* any such upturn.

Computer-aided manufacturing (CAM) refers to the use of computer systems to control the operations of a manufacturing plant. Referring back to Figure 6.5, these systems are directed towards the control of work analysis and planning and production control as well as tool and workpiece availability. In addition, CAM systems can be linked to CAD systems (CAD/CAM) where the output of the CAD system includes instructions for the manufacture of a designed product or component. This refers especially to CAD-generated programs for CNC machine tools. It can also cover other aspects of production organisation.

At present, few companies in small-batch engineering are anywhere near the construction of integrated CAD/CAM systems (Arnold and Senker, 1982). Where such systems do exist, such as in aerospace manufacture, the highly sophisticated machinery necessary is so expensive as to remove it completely from the ambit of most small and medium-size mechanical engineering establishments. In companies implementing CAM systems which are not linked to CAD, the problems of organisational structure become most pronounced. Such systems demand a well-ordered and rational product-flow system which can be modelled on a computer data-base. As one metal worker from a GEC factory put it: 'you can't computerise chaos' (Manchester Employment Research Group, 1983). In such cases, the necessary organisational changes, which have been ignored in favour of technological solutions to the costs of small-batch production reveal themselves to managements in the form of production bottlenecks.

The problems involved in the informational reorganisation associ-ated with CAM are crucially dependent on the difficulties of 'data capture', that is, linking the *model* of the production process in the computer system to the *real* process of production on the shop-floor. Whereas computer systems are extremely adept at the storage and manipulation of data, this is only effective where there is a continual updating of the model by means of the interface between computer and shop-floor. The use and reliability of the information system is ultimately dependent on the quality and speed of information which it receives. Hence, unless automatic forms of work and machine-function monitoring are devised, along with automated quality control systems, the functioning of the CAM system is very much dependent on *manual* updating of the database. Hence, despite the valid concern expressed by labour-process writers with the potential of information systems for monitoring the work pace of workers (CSE Micro-electronics Group, 1980) the phenomenon is double-edged. The greater threats of such systems, in fact, are to the jobs of supervisory workers carrying out production planning functions, as well as those of progress chasers and time-study people whose functions the information system can usurp.

Although in an early phase of development a particular attraction which CAM systems offer is harnessing the knowledge and experi-ence of shop-floor workers in the day-to-day running of the firm. The latter situation is the hope which Friedman (1977) discerned in responsible autonomy strategies on the part of 'enlightened manage-ment'. The compatibility of information systems co-ordinating semi-autonomous groups has been noted earlier as a strand in progressive management thought. The extent to which information systems associated with CAM are structured to enable managements to monitor crucial parameters of the overall 'decentralised' production process is paramount here.

CONCLUSION

What general conclusions can be drawn from this overview of technological and organisational developments in relation to the transformation of small batch engineering? First, it is apparent that whilst managers and engineers might wish to direct their efforts towards integrated, 'workerless' systems, the problems of integration founder on the partial nature of 'automation' in the individual components. So, problems of reliability arise from the lack of adequate means of monitoring and compensating for tool wear, for example in NC metal-cutting machine tools. Adding on flexible transfer systems can do little to resolve this problem. The result can be unreliability and extremely expensive systems lying idle for long periods.

Similarly, there are also the organisational problems of catering for modifications in production plans and changes in product mixes within highly mechanised systems, whose degree of flexibility may be high but not infinite. The degree of control over production (let alone over workers) which such informational systems can provide is therefore curtailed. These systems must thus be recognised as indicative rather than controlling in nature. Their viability relies ultimately on the quality of information supplied to the system both by machines and workers and the usefulness of the new information generated, whether in the form of a production schedule or a machine program.

It has been suggested that the end result of these attempts to transform small batch engineering into CAP systems is likely to differ from the grandiose claims of some of its most futurist protagonists. This is not to deny that a high degree of technical change is probable, but the likelihood of the fully mechanised, 'workerless' factory being the end-point of that change is similar to that of Group Technology forming the organisational answer to productivity bottlenecks in small-batch engineering. The practical result of the 'vision' will be different from the intentions of its promoters. In the case of small-batch engineering, rather than a centralised, highly mechanised system, the result is likely to take the form of tighter organisational control over production, but with workers exercising a relatively high degree of autonomy and skill at a local level. As the capital intensity of CAP systems is high, the premium for skilled labour is marginal compared to the cost of inefficient usage of capital equipment. So, rather than cutting labour costs by deskilling, as in Fordism, the inducement in neo-Fordist batch production systems is the efficiency of capital utilisation. This is aided by skilled operators more than unskilled machine minders.

These comments may go against the stream of engineering orthodoxy if one judges by articles in the technical press, but they reflect the fact that many of the 'workerless' factories are experimental, pilot or showpiece systems and rely on large amounts of state funding, often from military sources (see *Financial Times*, 3 November 1983). Nevertheless, taking a longer-term view, explorations of the technologies for such factories, whatever success they have as whole systems, will result in the development of partial technological and organisational competences. Subsequently, these can be successfully incorporated into smaller and simpler systems, where the levels of mechanisation and degrees of integration are lower. That there are economic and technological problems with full systems does not mean that small batch engineering is forever immune from productivity increases.[5]

To conclude: in this chapter we have reviewed various developments, both technological and organisational, in small-batch engineering and discussed how the problems of a straightforward application of Fordist principles of production organisation to that sector are being transcended by recent technological developments in the 'informational'/'control' dimension of mechanisation. We showed that the technological, organisational and informational principles on which production processes in small-batch engineering are being based marks a break with Fordism as we have defined it. By extension, in order to realise the potential of information technologies, to ensure an historic increase in productivity in small-batch capital goods manufacture and, in turn, contribute to a long-term upswing in economic growth in capitalist economies, would take a successful combination of neo-Fordist production process elements, the technological, the work organisational and the informational. We return to this in Chapter 8.

7 Information Technologies, the Service Sector and the Restructuring of Consumption

INTRODUCTION

As we first outlined in Chapter 2, information technologies are applicable not only to such manufacturing sectors as small-batch engineering but also to other sectors which have traditionally proved very difficult to organise on Fordist lines. In particular, this includes those sectors which have not been based on the use of much sophisticated machinery, either in terms of transformation, transfer or control. We refer to the application of information technologies in the supply and production of so-called 'services' to which this chapter is devoted. The chapter examines this sector to see whether neo-Fordist developments can be identified. After looking at what the service sector actually comprises, we examine various organisational changes that have taken place in one part of the sector that has been most affected by developments in computer technology since the 1950s – clerical work. We wish to develop the idea that information technologies being introduced into such work in the 1980s are best seen as associated with neo-Fordist organisational changes.

We will also seek to demonstrate that the concept of neo-Fordism cannot be confined merely to describing some transition *within* existing service industries or occupational sectors. Indeed the promise of information technologies, their recasting of traditionally labour-intensive activities of information handling, implies a substantial reorganisation of the existing service industries themselves, as the patterns of service consumption amongst the population alters. Just as Fordism for its extension to a large number of industries, required a reorganisation of the consumption patterns of the majority of the population so, we would argue, might neo-Fordism, though not necessarily of the same kind nor on the same scale. In other words, for a successful capitalist exploitation of information technologies, fairly substantial changes may be required in the way people in developed capitalist countries satisfy their needs. In the second half of this chapter we discuss this restructuring of consumption using possible changes in the ways in which health care is provided as an example.

THE SERVICE SECTOR

The service industries provide many of the prerequisites for the functioning and growth of capitalist economies. Many are concerned with the distribution of goods and raw materials, the circulation of capital, the production and reproduction of labour power and the provision of information and expertise. Many of these service industries are particularly labour-intensive and in all developed countries a large proportion of services are provided by the state rather than by private capital. Much of the non-state-financed service industries consist to a large extent of small capital and the self-employed. In contrast to manufacturing industries, few of the service industries have significant transnational capital, although this is beginning to change.

As the current economic crisis has continued, reductions in state expenditure on service industries along with the search – especially amongst equipment manufacturers – for new markets, has made the service industries very attractive targets for capitalisation. Similarly, the current contraction of the private manufacturing sector in many countries has forced service contractors to look to the state sector for growth especially in such areas as catering, cleaning and security.

The services do not clearly exist as a sector; they are a heterogeneous collection of industries. The main thing they have in common is that their 'products' are not in the main physical goods but are in some way intangible, impermanent and immaterial. A lot of ink has been spilt in attempting some embracing definition of what constitutes the service sector, much of it wasted by trying to bring unity to what is better left as diversity.[1] Gershuny and Miles (1983) have pointed out that the definitions are of limited value since the term 'services' has at least four meanings.

There are certainly *service industries* which provide the immaterial service products as well as physical products which have been transformed by such service activity. Such industries include the familiar, and sometimes overlapping, categories of tertiary industries (transport, communications, utilities), personal services (hairdressers, dentists, catering etc.), goods services (maintenance of cars, consumer goods and buildings, etc.), producer services (finance, banking, legal and research work), cultural industries (publishing, broadcasting, advertising, etc.) and public services (education, health and public administration). But many of the *service products* of these industries are also produced by manufacturing industries who offer them for sale (for example, computer manufacturers who offer maintenance services) or who consume them internally to their own organisation (for example, some firms have their own transport departments). So, *service workers*, people who actually produce the service products, are distributed across all industries, 'service' and manufacturing. In addition, it is necessary to identify *service functions*, necessary human requirements, or 'needs', which can be satisfied in a number of historically specific and changing ways. (For example, the human need to move from place to place can be satisfied by the transport *service industry* of buses and trains or by the self-drive physical products of the automobile manufacturing industry).

Exactly how the changing mode of satisfaction of various human service functions is linked to changes in the structure of service industries and employment is an important question. We will return to it later in considering the applications of information technologies. For the moment, let us note that Gershuny and Miles divide services into a number of categories thus:

1. MARKETED SERVICES – namely service products which are provided in advanced capitalist countries predominantly by private capital via competitive markets. These include:
   (a) **Producer services** – provided in the main to and for the capitalist- and state-production sectors rather than to individual consumers; namely:
   Finance – banking, credit, insurance, real estate, etc.
   Professional – legal, research, advertising, etc.
   Other business – cleaning, security, maintenance, etc.

(b) **Distributive services** provided to and for capital- and state-organisations for the transfer and storage of people, goods and information of all kinds; namely:
Transport, storage – the physical movement and storage of goods and people by road, rail, air and water
Communications – the physical and electronic transfer of letters, speech, financial and other data by postal, telephone and computerised services
Wholesale and retail trades – that is, the sale and trading of manufactured goods and of personal services

(c) **Personal services** – the provision of services direct to households or individuals or to households and individuals' possessions; namely:
Domestic – the performance of domestic activities such as cleaning, laundry, personal care (for example, hairdressing)
Lodging and catering – the provision of hotels and restaurant services
Repairs – the servicing of household durables and of domestic buildings
Entertainment and recreation – the provision of sports, arts, and general leisure/cultural services outside and, by cable, etc., inside the household

2. NON-MARKETED SERVICES – namely service products which tend to be provided by state-owned, state-financed and/or state-administered organisations in advanced capitalist countries; namely:
Health care – the services of doctors, other hospital workers, dentists, etc. within 'health services'
Education – education and training services offered by nurseries, schools, and higher and further educational institutes
Other welfare – personal social services (for example, social work), surveillance (policing) and other general government services (for example, public administration, communal cleaning).

We cannot go into the likely significance of innovations in information technology in all these service activities. Table 7.1 presents a summary of some of the most probable developments.

The concern of many writers to chart and quantify the growth of the services over the post-war period reflects their view that developed countries are moving to a 'post-manufacturing' or, less precisely, 'post-industrial' phase of development. In this phase, it is

150

TABLE 7.1   *New technologies and process and organisational innovations in services*

MARKETED SERVICES

PRODUCER SERVICES

Finance

Process Innovations: Electronic data processing for inter- and intra-organisation transfers; automatic dispensers and tellers; point of sale terminals; office automation on a large scale; use of terminals by agents in insurance and real estate, speeding up transactions.
Organisational Innovations: Specialised consultancy and accountancy services brokerage between different service institutions; movement of financial organisations into providing infrastructure for trade, small businesses and retail.

Professional

Process Innovations: Computer-aided design and routinisation of production of documentation with office automation; retrieval of data from extensive bases; some development of home-based office.
Organisational Innovations: Self-help and citizens' advice for legal and related services to private individuals extended with 'telematics'.

Other Business

Process Innovations: Telecommunications used to improve scheduling of tasks; greater use of automated security equipment, fast food and vending equipment.
Organisational Innovations: Contracting out of businesses' 'peripheral' tasks; increase in 'voluntary' (i.e. unpaid) work.

DISTRIBUTIVE SERVICES

Transport/Storage

Process Innovations: Automation of stock handling and accounting; more containerisation of transported goods; increased efficiency of rail transport with more rapid trains and improved routeing; use of telecommunications to improve booking, ticket sales automation, etc.
Organisational Innovations: In goods transport, private road transport services displacing rail; in passenger transport continued reliance on private motor-car with some increase in 'community transport' services; development of local goods delivery centres for final consumers in urban areas.

Communications

Process Innovations: Electronic handling of routine letters and packages; development of electronic mail, high capacity optical cables, more highly automatic telephone systems, communicating computers and word processors, access to data banks, videotext.

Organisational Innovations: New information services, complementing expansion of information and entertainment services for households; growth of community presses (and other media?) as 'alternative' information sources?

Wholesale/Retail

Process Innovations: Automated stocktaking, tied in retail sector to point of sales equipment and potentially linked directly to financial and wholesale institutions; growth of mail order complemented by 'teleshopping'; increased size of retail and wholesale units with more self-service.

Organisational Innovations: Delivery services tied to teleshopping; increased electronic consumer information systems; increased use of freezers and other domestic storage equipment with self-production expanding consumption of garden and household tools.

PERSONAL SERVICES

Domestic

Process Innovations: Increased productivity with economies of scale and specialisation in laundry and similar household services.

Organisational Innovations: Improved domestic technology – vacuum cleaners, washing machines, etc. – facilitates shift toward informal production of domestic services; flexible working hours plus women's movement pressure alters domestic sexual division of labour? Some professional services (such as computer software, interior design) offered direct to households via advanced telecommunications systems/home computers.

Lodging/Catering

Process Innovations: Improved reservation and booking services in hotel services; growth of 'fast food' outlets with off-site prepared food and microwave ovens, etc.

Organisational Innovations: Changing household patterns stimulates more 'eating out'; growing home use of pre-cooked, convenience foods.

Repairs

Process Innovations: Use of automatic diagnostic equipment spreading from car repair and household appliance maintenance; simplification of maintenance through use of microelectronic components.

Organisational Innovations: More do-it-yourself repairs subject to degree of 'penetrability' of electronic equipment.

Entertainment/Recreation

Process Innovations: More use of telecommunications in booking and sales; more 'hi-tech' entertainments.

Organisational Innovations: Growth of cable TV, video recorders displacing 'live' cinema and theatre further; extension of local TV and radio transmissions.

NON-MARKETED SERVICES

Health

Process Innovations: Computerisation of records and office automation; more efficient catering services; electronic diagnostic and monitoring equipment; new prosthetics technology; new ranges of pharmaceutical products.
Organisational Innovations: New drugs and biological diagnostic methods for various medical conditions; preventive and simple diagnostic services possibly made easier to use with information technologies; growth of paramedical and advice services; wide use of automatic diagnosis and screening in health services?

Education

Process Innovations: Distance learning systems enhanced through telecommunications, video and computer equipment; use of computers in 2° and 3° education.
Organisational Innovations: New educational packages for community education and for disadvantaged groups; educational packages combined with entertainment for home use.

Welfare

Process Innovations: Computerisation of records; automation of advice services and administration.
Organisational Innovations: Development of self-help groups producing community services and seeking improved services from state; blurring of policing/community work/surveillance roles in state agencies?

SOURCE *Adapted from Gershuny and Miles* (1983)

said, the dominance of manufacturing employment becomes more and more eclipsed by the growth of the service sector as the major source of employment. This is seen historically to mirror the decline of employment in primary industries of agriculture and extraction. However, the post-industrial thesis of the early and mid-1970s has been challenged by events in more recent times. The idea that manufacturing would become almost completely mechanised, leaving labour-intensive work in the service industries or (if much less so) in the service occupations of manufacturing, to take up the slack, assumed a low level of labour-saving technological change in services.

(This low level may have been due to the sheer technical difficulty of mechanising service tasks, the low pay of most service workers or the deliberate job-creating policies of inverventionist governments in the public sector.) It is clear now that this is not in fact likely to be the case, both because new technological developments can indeed mechanise many service tasks and because the prevailing economic and political situation discourages an extension of public, labour-absorbing services.

Yet while the optimistic social commentators of post-industrialism were replacing their vision of the 'service society' by a rather similar notion of an *information society* (this being a society where information is the dominant commodity being traded rather than manufactured goods)[2] other, pessimistic commentators foresaw the effects of information technology in services leading to a major shake-out of service workers in the so-called white-collar jobs (Jenkins and Sherman, 1979; Hines and Searle, 1979). The common thrust of these arguments was that there was likely to be a large degree of surplus labour – engaging either in 'leisure' or unemployment – depending on such matters as the policies of both the state and individual firms on job-sharing and, more vaguely, changing attitudes to the presumed 'work ethic'. The rising numbers of unemployed in developed societies from the late 1970s seemed to indicate confirmation of these trends.

Although few people would argue now that the high level of unemployment in the advanced capitalist countries is a direct result of the introduction of new technology in manufacturing and services, the present orthodoxy is that any upturn in the economy would result in jobless growth certainly in manufacturing and, most probably, in many services. While much of this writing has been the stimulus to debates on the effects of new technology, few commentators have been sensitive to the constraints of political economy, the barriers to increased mechanisation especially in the services, or the emerging patterns of technological and organisational change both within and between sectors.

Of course there are enormous difficulties in discerning technological and social trends at the level of the service sector as a whole just as it is difficult to talk about general trends in manufacturing as a whole rather than in specific industries. Yet as soon as we move from the aggregate to the level of particular industries and types of service, the multiple dynamics of service industries become easier to see. Recent work (Robertson *et al.*, 1982; Gershuny, 1983; Gershuny and Miles,

1983) has examined the main locations of growth in service industries. Over the post-war period around two-thirds of service industry growth is accounted for by the growth of state services, especially administration, education, welfare and, except in the case of the USA where it is mostly privately run, medicine. Most of the other one-third of post-war growth is accounted for by the expansion of producer services. This expansion is, in part, a reflection of the increasingly complex social division of labour in the production of goods and of the increasing importance of non-direct, conceptual tasks in contemporary manufacturing. (From 1924 to 1975, the proportion of administrative, technical and clerical workers in British manufacturing industry went up from just over 10 per cent to about 28 per cent; see Crum and Gudgen, 1978.) In aggregate terms there has been a relatively stable or slightly declining level of employment in the marketed services in developed capitalist societies. Especially in the cases of entertainment, transport and domestic services, growth in consumption in these areas to satisfy these service functions has been achieved through the purchase of goods like cars, television and domestic appliances.

As we have said, another way of characterising this service work is in terms of occupational categories. This has the advantage of capturing service work which does not take place in service industries as such. By defining service industries as concerned with the production of intangible products, as distinct from manufacturing production, such a definition neglects the dependence of a sizeable part of manufacturing activity on a variety of indirect servicing tasks, often, but not always, taking place in offices. These are an essential component of the production of the physical product: in Chapter 6, in the case of small-batch engineering, we described the work of design, planning, stock control, cost-accounting and progress-chasing which are non-direct production activities but are nevertheless essential to overall production. Increasing in importance are other white-collar functions within the manufacturing sector: marketing, research and development (R & D), purchasing, distribution, corporate finance and administration. This is in addition to indirect services such as plant and building maintenance, catering and cleaning. In this sense, a more adequate coverage of services is gleaned from service occupations rather than service industries despite the current trend for many services to be hived off or subcontracted from service firms.

In the rest of this chapter then we seek to focus on some specific service industries and occupations to illustrate some of the developments within which information technologies might be implicated.

CLERICAL INDUSTRIES

Information Technologies in the Office

Office machinery and computers, along with the telecommunications systems and associated software needed to operate these elements and link them together, collectively known as 'information technology' (IT), are certainly of great applicability to all service sectors. Indeed for many, IT will turn those services from highly labour-intensive, paper-shifting, minimal technology activities into fully-fledged tertiary mechanised industries, with massive leaps in labour productivity in a comparatively short period of time – hence the analogies which some writers draw with the Industrial Revolution which involved similar leaps for some manufacturing industries, cotton spinning in particular.

Consider, for example, the simple process of composing, typing and sending off a letter, say from some headquarter order department to a customer in response to an enquiry. Envisaged as a sequential process, the activity can be portrayed as in Figure 7.1.

The squares represent transformation processes in which, by hand and brain, words are conveyed to paper either for the first time or in correction. The various paper drafts are transferred, between author and typist and clerical assistant, entirely by hand up to the postal stage where their transfer to the recipient may be accomplished by mechanical means, either in sorting or in transport. Until recently, the technological component of these clerical labour processes was limited to dictaphones, typewriters, franking machines, post office sorting machines and motor and rail vehicles. Electronic office machinery and advanced telecommunications makes it possible (although over what timescale is disputed) to eliminate many of these steps in transformation and in transfer, as well as in some aspects of the control of transformation (for example, layout, pagination) embodied in the actions of skilled typists, by 'electronicising' the entire process after the initial data entry involved in typing. With automatic voice transcription devices, it has been suggested, there can be further reduction as even the necessity for typing is removed.[3]

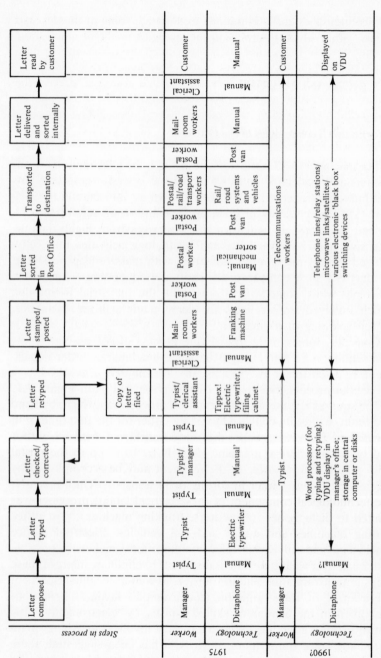

FIGURE 7.1 *The clerical labour process and new technologies*

The manipulation of the text of the letter, its permanent storage and its transfer to the customer become highly mechanised in all three dimensions with few places left for any need of human intervention after the initial composing and data entry actions.

Although such equipment is being installed at a brisk pace in the offices of some companies, the rate of diffusion must not be exaggerated. Office equipment manufacturers, as might be expected, forecast that fully integrated systems – the so-called electronic office – will rapidly diffuse into the larger offices over the next five years; but, as we pointed out in Chapter 2, the speed of diffusion depends on many factors, most particularly the presence of an adequate advanced telecommunications infrastructure to handle what would be a huge increase in the electronic transmission of words, data and graphics. Such infrastructures are being installed on a world-wide level and they make up the technological core of current political disputes over European state telecommunications authorities which are being dismembered in order to clear the way for international corporations to take over the long-distance national and international trunk lines. In contrast to the local telecommunications infrastructure these are very profitable and are especially used by large, often transnational, business users for corporate communication and data flows. The enormous problems of infrastructural provision (such as, who pays for cable TV? how can privacy be protected?) and equipment incompatibility beween systems of rival producers means that the arrival of the electronic office is not as imminent as the manufacturers would have us believe. A further problem arises from the advantages of cost, durability and portability which paper has over electronic communications equipment, as well as its being the already existing medium. The information technology 'poor' will use paper for a long time to come, hence it will be a rare office that turns out to be 'paperless'. In addition, even if some parts of the clerical production process are mechanised, the initial *preparation* of documents in the creative act of composition will still be a highly skilled, professional 'craft'.

Many commentators have seen the introduction of computerised office machinery – particularly the electronic word-processor – as the latest and most substantial extension of a long process of introducing Fordism into office work. In the rest of this section we examine this view. We shall argue that straightforward Fordist reorganisation of clerical work around information technologies will be confined (as it has been in the past) to those types of clerical work where standardisation of the clerical product is easiest. For clerical activities of high variety however – and it is these which have the lowest labour productivity – information-handling devices like word processors and telecommunication networks, although permitting some standardisation and reorganisation of work structures, are more limited in their effects and in this respect must be examined in different terms. As with our analysis of small-batch engineering developments we prefer to see the differences as constituting a different form of production process, namely neo-Fordism. As defined in Chapter 5, neo-Fordism is intended to describe processes displaying technological developments in control mechanisation which permit high variety in their transformative capacity, organisational changes which favour less hierarchical work-roles and an informational infrastructure which can integrate different sub-units of the production process to permit various forms of decentralisation. The information technologies that can be employed in offices favour such informational infrastructures since they permit the rapid on-line manipulation of all sorts of data.

Fordism in Office Work

(This section and the one immediately following draw extensively on Softley, 1984.) Offices of varying sizes are found in all branches of production but it has been in service industries with large amounts of clerical work, indeed where the service *is* the processing of all forms of paper (orders/receipts/money) where the most successful attempts to emulate Fordist forms of manufacturing organisation have taken place. The best examples of this are banking, insurance and mail-order administration. In his discussion of the history of office work, Braverman describes its reorganisation over the last sixty years as leading to the creation of: 'A stream of paper . . . which is processed in a continuous flow like that of the cannery, by workers organised in much the same way' (1974, p. 301). Strassman (1982) refers to this as the 'classical organisational structure' in which 'standardised management, using standard operations concepts, imposes standard operating procedures on its employees' to satisfy customers with 'standard tastes and standard purchasing habits'.

Such standardisation has been underway since the turn of this century. As offices grew in size, efficient decision-making processes became more and more crucial to the successful functioning of the enterprise. Rationalisation served to centralise this decision-making, which became increasingly identified with the management role. Fordist techniques, although first applied in manufacturing, began in the 1930s to be emulated in office work. Standardisation of work procedures, the reorganisation of office layout more on flow-line principles, and time and motion studies, were introduced as a means of increasing efficiency and productivity. Planning and co-ordination became centralised, thus removing decision-making from the lower levels of clerical work and forming the technical basis of management authority. Thus this attempt at Fordist rationalisation went hand in hand with the growth of the management role, even though levels of mechanisation were very low, being restricted to the typewriter, telephone and dictating machines.

This rationalisation partly resulted in the division between typing and secretarial work through the formation of typing pools. In Britain typing pools existed as early as the late 1880s, although it was not until the dictaphone came into widespread use in the 1920s that typing pools became the most common working arrangement for typists (Benet, 1972). A dictaphone allowed the physical separation of those who dictate from those who carry out the transcription from tapes, the audio-typist. Centralisation of the typing function within the organisation was believed to increase overall productivity, particularly that of the typists themselves.

Of course there is a considerable *variety* in clerical work in which each document may require some different processing. The low machine intensity remained until the 1960s when xerography and computers began to come into routine use.[4] These factors limited the applicability of Fordist standardisation techniques for increasing the productivity of office work over and above that gained by increasing the intensity of work through application of the division of labour. Attempts to standardise office work were therefore limited in their extent; Braverman is forced to point this out, even against the thrust of his own argument:

> The work processes of most offices are readily recognisable in industrial terms as continuous flow processes. In the main they consist of a flow of documents required to effect and record commercial transactions, contractual arrangements, etc. *While the processes are punctuated by personal interviews and correspondence, these serve to facilitate the flow of documentation.* (Braverman, 1974, p. 312, emphasis added)

The relative applicability of Fordist standardisation is plainly linked to the type of office work in question. The most successful types are undoubtedly the clerical industries. As has been said, in these offices the major activity is the processing of large volumes of information, often in a limited range of formats. So the major technique for the standardisation of inputs – buffering the internal bureaucratic systems and clerical labour process – is the use of forms. Application forms, order forms, cheques, etc., are all a result of the demand for an input in a format which will match the standard operations of the clerical labour process.

An addition to standardisation as a means of increasing the productivity of office work, although coming much later, has been the introduction of mainframe computers for electronic data processing (EDP) which became a commercial reality for most office work during the 1960s. In 1965 eighty-eight local authorities in Britain had access to a computer for administration. By 1976 the number had reached 260. Housing records, payrolls, accounts, engineering and architectural data and police records were put on computers. In banking, EDP lowered the cost of work, reduced labour levels and the high error rate, and, importantly, allowed more effective managerial control. Leavitt and Whisler (1958) had predicted that computers, by increasing the speed and quality of information from low to top management, would weaken the power and authority of middle management. Control of the organisation would thus be centralised. Several later empirical studies supported this view of Dale's:[5]

> As the machines [computers] increase in number and complexity, they give rise to further division of labour to add to those which, originally, they were called in to implement. Their costs rise and as institutions become larger there is a strong tendency towards centralisation; the physical aspect of the office becomes more like that of a factory department and the nature of the routine changes from purely clerical to the manual or technical. (Dale, 1962, p. 81)

However, there have been other views. Withington (1969) argued that computerisation only resulted in greater centralisation in routine areas, such as data processing, and that the tendency toward increasingly complex managerial heirarchical structures was not observed. Instead, some argued that computerisation would promote decentralisation of line responsibility (Likert, 1961). Individual offices that had an on-site computer, rather than having on-line access to the mainframe at head office, it was observed, had a greater local management autonomy and less centralised corporate authority. So, as the price of computers came down in the 1970s, it was suggested that as many more plants acquired their own computer facilities this would 'not lead to centralisation but foster greater structural complexity and more autonomy at the plant level' (Blau *et al.*, 1976, p. 37).

These accounts of decentralisation tend to confuse the notion of decentralisation of decision-making over an extremely narrow range of issues with the relocation of power and control over the way an organisation is run. As Crompton and Reid (1982) point out, 'decision-making does not of itself necessarily provide a reliable indicator of the location of power'. Instead, they argue, it is the 'capacity to influence or determine the premises upon which decisions are made' which is a more reliable indication of centralisation or decentralisation of power in an organisation. Thus, those studies that argue that EDP-type computerisation has promoted multi-level hierarchies and decentralisation of authority, neglect the fact that *decentralisation* of some decision-making functions may coexist with increasing *centralisation* of other, more strategic, functions. Although various middle managers may be able to take more responsbility for decisions through the availability of more information they do so within pre-determined premises, that is within limits that have been set by higher management.

Thus, computerisation radically affected many jobs in the office and altered its structure and dynamics. Unlike earlier forms of mechanisation, computerisation had a profound effect on the organisational structure. It also required a reappraisal of both data and information flow. The application of mathematical techniques in conjunction with electronic data processing and operational research rationalised decision-making. This directly affected planning procedures and management and often led to recentralisation as many previously separate departments (such as sales and marketing) were reintegrated into a centralised office structure.

In the office, those adversely affected by computerisation were principally the female clerical workers who, with the use of EDP, became key-punch operators. Key-punching became the largest single category of computer-related work, contradicting claims that computerisation would immediately mean a reduction in the number of lower skill jobs (Shepard, 1971). Hoos (1961) studied the impact of computerisation upon female clerks in the USA and found that former filing and ledger clerks, of whom the vast majority were women, were subsequently employed as key-punch operators. According to Hoos, this work was monotonous and routine, it paid less than clerical work, and there was little chance of promotion. The other jobs created by computerisation – programming and systems analysis – required special training which the majority of women clerical workers would be unable to acquire. Lack of educational qualifications and the responsibility of married women for household tasks and child-care both contributed to the difficulties for women attempting to take advantage of such positions. Further, key-punch work was more easily controlled and measured by office managers than other clerical work. So in conclusion, Hoos argued that there was little evidence to support the claim that what she called 'automation' provided work of greater interest for the clerical workers. Thus it was the bottom group of the office hierarchy which benefited least from the reclassification of office jobs and which today continues to fill those jobs with the lowest status and lowest pay. That bottom group remains almost exclusively female.

There seems little doubt then that some principles of Fordism were applicable to computer-dependent clerical processes, contrary to those who associated computerisation with decentralisation and upgrading. But these early applications of computerised data processing were limited by two factors. The first was that only highly structured operations, such as accounts, wages or administrative functions which could be represented in terms of numerical data, could be easily computerised. The second was the high cost of using mainframe computers, which necessitated the processing of high volumes of data in centralised computer departments.

On the other hand the application of Fordist principles to the highly *varied* work of most offices was not an easy one. In such offices, both in services and manufacturing, which exist outside the large-scale clerical industries, the scope for routinisation and standardisation of procedures has been limited. This is partly due to the smaller scale of office work, which has limited the sub-division of clerical activities such as in typing pools or filing departments. Also, since much of the work is based around *particular* customer requirements (as in the case of service offices) or a physical production process (in the case of manufacturing) the scope for Fordist rationalisation is more limited than in some of the 'clerical factories'. Although office procedures do exist for activities such as ordering and invoicing, stock control and cost-accounting, a significant proportion of routinised office work is concerned with attempting to fit the information to be processed *into* the procedures, and with subsequent 'trouble-shooting' activities. Thus, although management, operations and customer requirements, in contrast to the 'classical' organisational structure, may be in Strassman's (1982) term 'diverse', the *procedures* to be followed will be standardised. This is particularly the case in retail offices such as estate agents, travel agents, professionals' offices, which have shop-fronts, where the requirements of individual customers or clients have to be dealt with. Where service industries are more highly concentrated, these smaller offices are often satellites of the larger clerical headquarters. As such they can be seen as buffers between customers or clients and the larger organisations – as in the case of the local office of a bank or building society, for example.

Neo-Fordism in Offices

So if Fordism, dependent for its success as a form of production on standardisation, has had only a limited application within office work, what difference is made by the arrival of new computerised technologies of a much less centralised and fixed type than the mainframe computers of the 1960s? Despite the familiar claims that electronic word-processors (the most common piece of computerised new technology to be used in offices) would upgrade office workers by eliminating tedious routine tasks like the repeated retyping of corrected documents, it has been argued by some authors that such equipment has *intensified* prevailing trends, making Fordist forms of work organisation easier to install and increasing the division of labour and speed of work.[6]

A word-processing system can be described as a combination of equipment and personnel, working in an environment of job specialisation and supervisory controls, for the purpose of producing typed documents in a routinised, cost-effective manner. Given this classical Fordist description, it is unsurprising that little has been claimed, either by manufacturers or the management literature, regarding the enhancement of *typists'* work with the use of word-processors; this is in contrast to the favourable predictions concerning the effect that word-processors would have upon *secretarial* work. What is usually claimed by manufacturers of word-processing systems is that they make a typist's work easier. However, this must be offset by the effects of standardisation, fragmentation, increased work load and supervision.

When word-processors are introduced into a typing pool the overall effect is to rationalise clerical production further and to intensify work. This process begins with an initial work study and continues with the automatic monitoring and increased supervision of work. As Softley (1984) describes:

> Before a company decides to purchase a word-processing system it will usually undertake some form of work study, entailing detailed accounting of the work being carried out in the typing pool, e.g. the number and type of documents typed per day and the number of separate revisions per document. If the decision to purchase is made then the results of the work study are used to standardise the form in which the work reaches the centre.

For instance, an insurance company had to rationalise a hundred different types of forms so that a few standard layouts could be placed in the word-processor memory. Another insurance company failed to do this initially but soon realised that productivity increases depended on standardisation and began to alter the layout of the forms they used. If an organisation is intending to set up communications systems between its different branches, then a standard format in all offices would be even more necessary. Another consequence of standards however, is the control they give to supervisors and management over production- and work-scheduling. Standardisation of work also facilitates greater fragmentation of tasks. In shared-logic or distributed systems, where each terminal is linked to a central processing unit, work can be automatically transferred from one typist's terminal to another. A group of typists can then handle one long report simultaneously, thus increasing the division of labour in the pool. No longer do individual typists work through a piece of work but, depending on demand, they are required to work on parts of other documents without necessarily seeing the beginning or end of it, or knowing anything about what it is they are typing.

Nevertheless, standardisation of work does not necessarily mean a decrease in the variety of the work performed. The majority of word-processor operators in Softley's survey indicated that word-processing had increased the variety of their work. It could be argued that the incorporation of word-processing into typing systems is a relatively recent phenomenon so that any changes now being observed may be purely transitory in nature. Yet, this increase in variety has also been reported in the USA where the average length of time a company had used word-processing was five years compared with two in Britain. The observed increased variety of work can be best explained by reference to the way work is organised in a traditional typing pool where tasks are already standardised and fragmented. An increase in variety in the move from typing pool to word-processing centre may simply indicate that different types of work are now being carried out, although the rules and procedures for completing each category of work remain standardised.

Even though electronic word-processors seem to have exacerbated the Fordist aspects of typing pool work, it must be recalled that not all clerical workers are engaged in full-time typing so alternative forms of work organisation have to be envisaged. In the early 1970s when word-processing was first being introduced into the USA on a large scale the division of secretarial work into its typing and non-typing components, followed by centralisation of the typing/word-processing function, was indeed seen as the most cost-effective way to increase productivity. Discussion in management journals focused on the efficiency of word-processing centres that could handle the typing needs of the whole organisation. The typists, or word-processor operators, in these centres became known as 'correspondence' secretaries. The remaining secretarial support was provided by groups of 'administrative' secretaries, who carried out filing, telephoning and reception work.[7] Thus, the centralisation extended to the secretarial function. Sometimes secretaries would have typewriters for short memos, otherwise all material for typing would be sent to the word-processing centre.

In terms of the typing function, such centralisation represents a highly rationalised arrangement – the word-processing centre usually being administered by the office services department. In many companies, however, there has been considerable resistance to the idea of centralised control over word-processing departments. Individual departments and divisions often want to retain control over various aspects of the word-processing centre's work – in particular supervision and prioritisation of work – leaving the office services department responsible for system and job design, installation of the equipment and training. In the USA, there has been a move *away* from full-scale centralisation. Softley (1984) reports a company which, in the initial enthusiasm surrounding the introduction of word-processing ten years ago, set up a word-processing centre. It was soon discovered, however, that one centre did not provide enough flexibility. Problems arose with the scheduling of material and the long turnaround times, even for such things as notes and memos. A few years later, the secretaries, who had been grouped into administrative centres, were given back their typewriters and the correspondence centre was split into three smaller ones. The word-processing centre had been decentralised.

In Britain, the introduction of word-processing has not so far meant much large-scale organisational change. For the most part, companies have simply switched from centralised typing to centralised word-processing (Manpower Services Commission, 1982; Steffens, 1983). At the time of writing, word-processing is just beginning to have some impact on the work of British secretaries. In small to medium-sized companies the quantity of repetitive typing work carried out by personal secretaries does not usually justify the purchase of a word-processor and in other organisations secretaries have access to a word-processor but are not usually employed as full-time operators.

It must be remembered that electronic word-processors are only one component of computerised office equipment that is likely to be introduced into offices over the next decade. The other equipment – particularly desktop computers and intelligent telecommunications devices – will be significant in boosting the productivity of office workers who are not typists. As Strassman (1982), a senior executive with Xerox, the office equipment company, put it, '(Such equipment) can compress the time for completing work . . . by altogether eliminating many labour-intensive office tasks performed by craft specialists whose burdensome and sometimes frustrating jobs are now embedded into the electronic workstation by means of software'.

Just as with manufacturing industry, some writers have suggested that non-Fordist types of work organisation are more appropriate for the operation of these high variety computerised technologies if maximum efficiency is to be obtained from them. However, there have been fewer studies of such alternative forms of job design for office work than for manufacturing work. Mumford is one of the most prolific writers on this subject. She has carried out several studies of the introduction and design of computer systems (Mumford and Henshall, 1983; Mumford, 1983; Mumford, Land and Hawgood, 1978). Like socio-technical systems theorists (see Chapter 5) Mumford argues that the structure of workgroups need not be determined solely by the technological system, but that a fair degree of 'organisational choice' is possible, allowing the formation of non-specialised, but integrated, workgroups. Designing work systems in this way is supposed to increase job satisfaction for the workers concerned, and the involvement of workers in the actual design of work systems is a key feature of Mumford's approach. Most recently Mumford has been involved in the design of word-processing work systems in private companies, where unionisation among typists and secretaries is low.

Mumford uses the concept of a 'design group' in her work, meaning a participative approach to changing work methods which aims to balance the efficiency needs of the organisation against worker satisfaction. Systems should be designed which are not simply concerned with the choice of computer hardware but also with the overall organisational system, 'which includes a network of people carrying out a variety of roles and having individual and job satisfaction needs' (Mumford *et al.*, 1978, p. 2). The design process is therefore much more extensive than was envisaged when early training programmes for systems analysts were conceived: 'It requires of the system analyst an ability to identify and specify both organisational and social needs and to create a socio-technical system that facilitates the achievement of these needs' (ibid). The design process uses several of the standard analytical/design tools that have been traditionally used by socio-technical analysts, but by arguing for a group participative approach to the design, evaluation and implementation of new systems, Mumford departs from the standard way of designing work systems.

The design process has been applied by Mumford in the redesign of a system at ICI's central management division. A video, *Reliving the Journey*, has been made of the process. Between them the ICI secretaries work for over fifty managers. With such a heavy workload, the word-processors that had been installed were in constant demand and problems arose with the scheduling of work. The video describes how the secretaries analysed their workloads and decided on the most appropriate work-scheduling process (Mumford, 1983).

In another project, involving clerical workers at Rolls Royce's Derby plant, Mumford set up a programme of workers' involvement in the design of a new on-line computer system. A design group was set up composed of workers from the departments involved in the reorganisation, together with a steering group made up of middle and upper level management and a trade union official. After studying various alternatives, the design group drew up three system designs which were submitted for comments to both clerks and management. One version was then chosen and implemented. Kraft (1979) has criticised this approach because the clerks were not given the choice as to whether they wanted a change or not, the members of the design group were selected by management rather than being elected by their fellow clerks, no mention was made of issues such as job security and work loads and the trades union was relegated to only a minor role. In the event, management agreed to implement an alternative favoured by the clerks, but only as a short-term measure, for they intended to introduce their favoured system 'when the clerks had got used to the autonomous group concept' (Mumford and Henshall, 1983, p. 75).

Kraft's criticisms show that the libertarian, work-humanising ideologies of job redesigners like Mumford are severely constrained by the realities of management power and prerogative, but this is partly beside the point. What is important to managements eager to get the maximum productivity benefits out of new technology is that their workers should be sufficiently flexible in their approach to work-load scheduling. The reorganisation of work around 'autonomous work groups' is simply one means to this end. Of course, some employees and their representative trade unions may have objections to such new forms of work organisation, particularly if they imply a reduction in employment or some uncompensated increase in overall work-load or some alteration of demarcation lines. This is why job redesign, in office work as well as in manufacturing, is easier to carry out where trade unions are non-existent or weak or under strong pressure because of some unfavourable competitive situation in their firm.

From the viewpoint of managements, the successful economic introduction of information technologies requires some new forms of work organisation – one of the elements of neo-Fordism – and it is the opposition of workforces committed to the institutionalised divisions associated with existing 'craft' or Fordist labour processes that managements must seek to overcome. The job redesign procedures advocated by Mumford for office work based on computerised technologies are a means of seeking to engage the commitment of workers to more flexible working arrangements which must go with technologies that raise the productivity of high variety labour processes. Autonomous work-groups, implying the participation of workers in the design of their jobs and in the allocation of work tasks require a more wholehearted commitment to the overall production process than might be expected by the application of more coercive Fordist principles of labour division and task allocation associated with standardised work routines. Such a view is implicit in the idea that 'future administrative systems' using information technologies will possess:

> greater diversity in procedures that can readily be customised to respond to the need of diverse customers. This diversity in procedure is built up by combining standard modules of know-how, usually incorporated into software, together with an increased amount of choice made possible through individual initiative and through work group co-operation. (Strassman, 1982)

The use of autonomous work-groups is however not the only form of non-Fordist organisation that can be envisaged using information technologies to handle information. If close contact between members of the group is essential to the work they perform then group working will be particularly appropriate. But advanced telecommunications technologies make it possible for much clerical work to be done in a geographically dispersed manner with centralised monitoring and co-ordination of the 'remote workers' who will be 'teleworkers' working at computer terminals in their own homes.

Huws (1984) reports a number of examples of teleworking in Britain and the USA. Whereas most teleworkers at the moment tend to be higher professionals, often working freelance within the computer industry itself, there is a growing, if still small, number of non-professionals. Some are involved in 'offshore information processing'. American Airlines, for example, closed down its data entry operations in Tulsa, Oklahoma in 1983 and hired 200 Barbadians to do the work, using a satellite link between the Barbados workers and the Tulsa data-processing centre. (In the age of modern telecommunications, everywhere it seems is less than twenty-four hours from Tulsa! (Pitney, 1963)).

Given the prevailing sexual division of domestic labour, most of the teleworkers Huws surveyed were women, almost all with child-care responsibilities. Telework then, although convenient for otherwise housebound women who wish to work for wages, has many of the disadvantages of domestic outwork – isolation, unsocial working hours, inadequate technical support, no fringe benefits.

Such arrangements are however highly advantageous to the subcontracting firms. Huws reports:

> (One firm) estimates that homeworkers produce about 30 per cent more than office-based workers in the same time, while ICL claims that twenty-five hours work in the home is equivalent to forty in the office . . . Another important advantage of employing homeworkers was the reduction in overhead costs. These did not just consist of floorspace, heating and lighting, but also included administrative support services and, where homeworkers were self-employed, perks such as company cars, subsidised meals, BUPA and pension schemes as well as holiday, sickness and maternity pay. Flexibility of working hours was also cited as a positive feature of employing homeworkers. This was particularly important in the jobs which made major demands on central data-processing facilities, since work could be timetabled to fit in with times when there was spare capacity on the computers, and off-peak telecommunication charges applied. (Huws, 1984, p. 16)

But there are also a number of disadvantages – monitoring and supervision of non-routine homework is more difficult than when workers are on the organisation's premises. Obviously, telework's future is dependent on whether techniques of remote monitoring can be devised, as well as on the complex calculus, by firms and by potential teleworkers, of factors such as the relative costs of telecommunications, office rents, transport to work and wages.

Conclusion

To summarise this section, whereas it has been possible to transform the production of some clerical work services along Fordist lines, this has generally been limited by the degree of variety of the clerical work concerned. The success of such transformation and reorganisation has been strongest in those services which could be described as clerical industries. These include services like banking, insurance and finance, mail order and public administration. In general, other clerical work has proved much more difficult to reorganise along Fordist lines, and as yet, the computerised office technologies in use in offices have tended to be restricted to one part of the clerical process – typing. The integration of separate pieces of office equipment, within and between offices, and their diffusion into many more service sectors – in short the increased use of flexible office equipment – will permit all sorts of 'information workers' to perform a wider array of non-standardised tasks at higher productivity levels. This suggests that the limitations of Fordist applications might be transcended. This will be especially the case if new concepts of job design are widely adopted or if telework becomes more acceptable. Together one might characterise these changes as neo-Fordism in clerical work.

THE RESTRUCTURING OF SERVICE PRODUCTS

However, the application of such neo-Fordist principles to the organisation of clerical processes in the production of the services is not the sole way in which the limitations of Fordism can be overcome; indeed, although it is certainly important in the short term, we are of the view that there are much more substantial changes in service provision that can be envisaged. In other words, the services could also be transformed by changing the way in which they are delivered, that is by changing the form of the service product itself. This can be done in two ways: by *externalising* part of the service labour, shifting it on to the consumers themselves; and, more dynamically, by changes in the *kinds* of services which are offered to satisfy human needs. We elaborate on each of these in turn.

Externalisation of Service Labour

One way round the problems of significantly raising labour productivity in services is by the externalisation of the labour of providing the service. Thus, rather than provide the service 'in full', service production takes the form of providing the users with the means to provide the service, to varying degrees, for themselves. The user contributes some of the labour, unpaid, whilst capital provides the means of production! The externalisation of labour can take various forms. One form is that of reorganisation of service industries so as to incorporate a *self-service* element, familiar examples of this being in restaurants and supermarkets. Here, it is the equivalent of the *assembly operations of Fordism* which are externalised. Going around a supermarket, shoppers assemble their particular basket of goods from the module of components which are presented to them by waged service workers (shelf-fillers, stock controllers, etc.). Similarly, in a self-service restaurant a meal is 'assembled' from a choice of components. Other consumer services such as banks and libraries are adopting this technique, organising the self-service activity around information technology (such as cash points).

The transformation of the labour process in the delivery of services is probably most developed in the case of supermarket retailing. This process began in inter-war America where use of self-service spread horizontally to practically all types of retail stores; only later did these techniques diffuse through British retailing (Dawson, 1981, p. 22). The development of supermarkets in the USA during the 1930s depended not only on the provision of private transport and a developed road infrastructure to enable the volume of sales to be increased, but also on the availability of standardised products. The prepackaging of food by canning and wrapping, along with the use of refrigeration and preservatives to ensure shelf life and also the grading of perishables, allowed a wide variety of goods to be available from different suppliers. Yet externalisation of labour allows the sloughing off of more labour-intensive bespoke work, the assembly of different individual baskets of goods, on to the consumer. In retailing then, the servicing tasks of selling, weighing out, bagging and standardisation of products, are displaced from the retailer to the supplier, whilst the selection, retrieval of goods and delivery are shifted onto the buyer with self-service and cash and carry; so the economies of scale of supermarket retailing only capture part of the rising productivity of retailing, the other parts being accounted for by self-service, together with food-processing and packaging developments.

A similar case is that of catering, where the assembly line of the cafeteria resembles that of Fordist production processes. There are analogous problems of sequentiality and line-balancing just like the assembly lines in production, since the speed of the line is dictated by the slowest customer and bottlenecks often occur. Currently several of these cafeteria systems are being changed from one single line to several mini-lines for different parts of the meal (meat dishes, fish dishes, sweets, drinks, etc.) in an attempt to overcome such problems. Such reorganisation of flow-lines is similar to parallel flow reorganisation in production assembly, detailed in Chapter 6. Note that supermarkets do not face this problem since customers are not bound to follow a prearranged route through the supermarket; they can vary their route according to the desired constituents of their shopping baskets. In this sense the supermarket and reorganised cafeteria have more in common with neo-Fordist forms of organisation than Fordist ones. The exception to this is, of course, the check-out queue. It is to this bottleneck that computerised technologies such as electronic point-of-sales systems are being directed.[8]

The introduction of computerised information-handling systems makes externalisation possible in some other service industries. Most notably, the rapid diffusion of automatic tellers/cashpoints inside and outside banks can be seen as a way of increasing the productivity of bank workers by mechanising simple monetary transactions. One can envisage similar developments in such labour-intensive services as ticket sales, food and drink vending and non-complex information and advice dissemination.

The Restructuring of Consumption

An an alternative to externalisation, the means of production of services can be sold direct to the user in the form of a physical commodity. This has already been an aspect of the Fordist restructuring of consumption. Rather than producing the means of production for the collective provision of services, some industries have concentrated on the provision of consumer goods for individual or household consumption. To explore this further we will briefly recapitulate and extend some of the ideas of Aglietta presented earlier in Chapter 3.

Aglietta characterises the historical development of capitalism in the USA as a regime of extensive accumulation in the nineteenth century followed by a regime of intensive accumulation, which was constituted politically in the 1930s and 1940s in the New Deal and in World War Two and generalised world-wide after the war. But Aglietta's major contribution is his description of this shift as a 'simultaneous revolution of both the labour process and the conditions of existence of the wage-earning class' (Aglietta, 1979, p. 20), together constituting a Fordist regime of intensive accumulation. Aglietta therefore sees Fordism not only as a set of principles for organising a production process in a particular way but also as involving the fact that *consumption* of those products has to be 'organised' to guarantee profitable sales of the mass-produced items.[9]

This change in the 'conditions of existence of the working class' came about through the development of a homogenised mode of consumption based on the commodities from old and new branches of production organised along mass-production lines. The change took place in two phases. First, the depression of the 1930s is explained as a crisis in the extension of the norms of consumption from the rising 'middle class', who were the main markets for the growing new consumer industries of the early twentieth century, to the working class who so far had not significantly participated in purchasing these new products. The New Deal in the USA and the adoption of Keynesian welfare state policies in Europe after World War Two restructured capitalist economies from their pre-war depression to lay the basis for a post-war phase of energetic intensive accumulation. Second, the most recent sustained period of accumulation from the 1940s onwards involved the diffusion of Fordism throughout the mass-production sectors of capitalist economies. The potential profitability of this form of the labour process crucially depended on the active construction of economies of scale, and therefore on the restructing of consumption.

While we do not feel that Aglietta's account of how labour process change was linked to phases of economic growth is adequate by itself – indeed the first three chapters of this book constitute an account with a different emphasis – we do agree with him that Fordism implies an analysis of consumption as well as production. He sees the consumption changes as the following:

- *the commoditisation of the means of consumption:* the transformation of domestic household labour activities in such a way as to open up markets for profitable production of consumer goods aimed at the individual household. Examples would be factory-produced clothing instead of home-made; bought and semi-processed foods instead of domestically grown and prepared; the use of domestic appliances and automobiles, etc.
- *homogenisation of this consumption* by means of standardised mass-produced commodities. The economies which this form of production allows undermine the necessity for and economic viability of domestically-produced alternatives.
- *an extension of these consumption norms* to more and more sections of the working class in advanced countries, not least to non-working dependents (the old, the sick).

The developments in the 1930s and 1940s which Aglietta sees as facilitating these trends are:

- an infrastructure of housing, roads and utilities within which a large consumer-durable market could arise.
- collective bargaining, which permitted planned increases in wages and the reduction of working time, especially for white, male, organised workers and established income security for less-well-organised sections of workers.
- welfare systems which introduced a minimum social wage and assisted in the generalisation of the social consumption norm.
- financial innovations – credit, hire purchase – to facilitate purchases.

Thus, by the 1950s and early 1960s it had been possible to generalise Fordist labour processes, extending them to industries producing what were 'luxury' goods before the war. Both old products and new products fitted into the Fordist production norm which was aided by short-life, cheap goods (planned obsolescence) and the homogenisation of oligopolised markets by advertising.

Despite its prolonged success there are progressive limits to the continued efficiency of Fordism. Aglietta, like us, sees these limits as part of the present economic crisis. These exist both in established Fordist production processes, which we have already mentioned in Chapter 3, and in the extension of Fordism to other sectors.[10]

The importance of this analysis for our discussion of the service sector is its emphasis on how the nature and form of some service provision has already been radically changed by Fordism. This raises the question of how further changes in service provision might be brought about, either by development of Fordism or by neo-Fordism.

Aglietta draws attention to the limited applicability of Fordism to reducing the cost of the collective welfare services of health care, education and public administration. Echoing some British monetarist writers, Aglietta suggests that the low growth in productivity, but rise in real wages, in these labour-intensive service sectors has acted as a brake on accumulation. This can only be resolved by 'radically transforming the conditions of production of the means of collective consumption [the welfare services]', renewed accumulation only beginning with 'a massive transformation of unproductive labour into labour productive of surplus value' (Aglietta, 1979, p. 157).

A similar analysis of services, though not situated within the development of the organisation of production or within broader politico-economic factors relating to the progress of capital accumulation, has been put forward by Gershuny and Miles (1983). They argue that over the past fifty years major service functions have been affected by a change in the pattern of service consumption. The transport service function, for example, was satisfied in the 1930s by a transport service industry of railways, buses and trams, privately- and publicly-provided, which offered many people occupations in the transport service. Technologically, the industry combined these occupations with heavy machinery (the trains, the buses) within an established physical infrastructure of the railway and road systems. These combinations of technology, infrastructure and specific social arrangements Gershuny and Miles call 'socio-technical systems' (not to be confused with the identical term for a type of work organisation, as discussed in Chapter 5).

In the 1980s, however, there is a strikingly different socio-technical system of transport service function provision. Whereas collectively-used transport service industries are still in business (and now embrace air travel as well) there has been a shift to the satisfaction of this function by 'self-service' modes. Encouraged by the cheapness of mass-produced commodities, individuals now purchase hardware from the manufacturing sector (namely cars in this example) and drive them themselves, in effect using their own unpaid labour. The balance between the formal service industry provision of transport and its 'informal' self-service provision has shifted. The service function has become 'privatised' in the sense that every household is encouraged to purchase its own car. What follows is falling employment in the public transport service industry and, until recently, rising employment in the manufacture of transport products. Gershuny and Miles argue that similar shifts have occurred in the satisfaction of other service functions – particularly in the domestic, entertainment and communications service functions, which are now reorganised around manufactured commodities.

In entertainment, direct entertainment (theatre, cinema) has been replaced by broadcasting and receiving equipment and video- and audio-recording and playback equipment with associated software – TV programmes, records, video tapes. In the domestic services, the balance has shifted, notably with the provision of domestic appliances but also with 'do-it-yourself' equipment for other domestic services such as building maintenance, decorating and appliance and car repairs. In short, the parallel development of production and consumption based on Fordist industries has acted to expand and transform consumption of entertainment, transport and domestic services away from provision by service industries towards the consumption of goods.[11]

The transformations in these service sectors has been termed by Gershuny and Miles 'social innovation', based on the growth of what they call self-servicing (to be distinguished from the form of self-servicing described earlier as 'externalisation'). The penetration of service sectors and their transformation by, originally American, large capital has been indirect through the marketing of mass-consumption products rather than transformation of the existing service industries themselves.

A necessary condition of this development, as we have already pointed out, was the provision and extension of particular infrastructures on which the mass-produced commodities of cars, domestic appliances and electronic consumer goods were dependent; public roads and later motorways, the national utility networks, the broadcasting network. Of course, this domestic consumption was also linked to the increasing provision of adequate public and private housing over the post-war period.

Despite the relative decline of service industries providing equivalent functions to those of the new mass-produced commodities (public transport, domestic services, cinemas and theatres) a new type of service grew – which Gershuny and Miles have labelled 'intermediate consumer services'. Unlike the provision of service functions taking the form of final consumer services, intermediate consumer services involve servicing the new forms of consumption: for example, domestic appliance repair and maintenance, broadcasting stations, garages and recording industries. In the case of record and, later, video industries, this intermediate service involves the production and sale of goods, such as records, audio tapes, video tapes, as a physical vehicle for the 'software' service, as distinct from broadcasting which provides the software instantaneously rather than in the form of a material commodity.

One crucial difference between the purchase of services and the consumption of service-replacing consumer goods lies in the provision of labour needed to produce the final service. In the case of 'social innovation' the labour is ultimately supplied by the consumer. The 'labour-saving' effect of domestic appliances on domestic labour is a form of labour externalisation in service functions.[12] Rather than enter and transform the low-productivity traditional service sector of, for example, laundries, domestic decorating or repairs, passenger transport or live entertainment, capital investment took place instead in mass-production of consumer goods.

So the transformation of service functions by the use of Fordist products is complex. It fulfils certain needs (mobility, washing clothes, leisure, etc.) by means of a changed product but it also assists certain types of firms differentially because of the relative ease of increasing productivity and profits in, say, automobile engineering as compared to transport services. Also, the total level of demand for service functions is expanded by channelling consumers' needs into the profitable consumption of Fordist commodities, rather than the use of collective services. The problem of increasing the productivity of labour in service industries is, in effect, circumvented by externalising the necessary labour from the formal (paid) to the informal (unpaid) sector. The question now is: do information technologies make such developments more 'attractive' in other service sectors?

Restructuring State Services

The group of services least affected by the developments discussed so far are those provided by the state. Authors of many conflicting ideological persuasions agree that any long-term economic upswing in advanced capitalist economies – at any rate an upswing which maintains the essential features of the capitalist mode of production – is threatened by the large proportion of the national product which is spent on state-provided services.[13] These services may be thought of as welfare provisions, but they are also essential for profitable industry insofar as they generalise the costs of the 'reproduction' of the workforce. They provide education, training and health care for the workforce of today and tomorrow, and services like rubbish collection, sewage treatment, roads and energy supplies, which maintain general social health and offer cheap means of production and distribution for the output of private industry. As a result of the enormous social, health and environmental problems which the rapid industrial growth of the past thirty years has produced, other services have grown to deal with the diseconomies of capitalist production. Hence there has been the growth of regulatory services to deal with pollution and environmental problems, of planning services to attempt to bring some order to the otherwise anarchic locational decisions of modern corporations and of social work services to police and provide sustenance to those who do not quite fit the personality requirements of the modern good citizen. Added to this, for some countries, is the enormous cost of providing military arsenals to protect the perceived interests of the advanced capitalist states.

The emergence of the welfare state in Britain after the 'post-war settlement' involved an acknowledgement by the ruling class, via the state, of the inability of market mechanisms to fulfil certain social needs – particularly those associated with the reproduction of the labour force and the maintenance of the non-working population. It was a result both of class struggles for improved housing, health and education and of a perceived need for more instruments of social control (Gough, 1979, p. 52). The post-war Welfare State took on responsibilities for the provision of collective social services, which have in turn undergone a fairly continuous growth during the post-war boom. The growth of collective provision in education and in medicine contrasts with the reduced provision of those collective services providing parallel facilities in transport, entertainment and domestic services which, as we have already argued, were the main targets of the Fordist production of consumer goods.

In fact, the collective services of education, welfare and medicine are less amenable to commodity substitution than those of transport, entertainment and domestic work. They involve large elements of professional work which has acted as a brake on social innovation in the provision of such services. There are some examples of commoditised, self-service varieties, such as home instruction courses, community care schemes and patent medicines but these form the minority of service provision. A second function of such services provided by the state is that of ensuring social control and the adequate reproduction of the workforce. In the case of welfare services especially, but also in large parts of education and medicine, state responsibility reflects a broad concern with ensuring particular standards of child-care, socialisation and health. This often involves the support and policing of family units, these being the preferred site of social reproduction. Institutions of education and medicine have been developed so as to augment rather than replace the family as a fundamental site of reproduction. So total service provision (for example, children's homes or long-stay hospital wards) are seen as necessary only when family-based provision is declared inadequate by state functionaries. Hence the scope for self-service, commoditised replacement of state services has been limited by the need for the state to exercise social control functions via professional agencies.

As the economic situation has deteriorated, increasing pressure has been felt to restructure this pattern of state services in all advanced capitalist countries, though to different degrees. This has come from various sources, which include:

- political parties and parts of the state itself. Monetarist strategies to reduce the overall level of state expenditure have resulted in cuts in state services and attempts to increase the efficiency of welfare services like medicine, social services, education, cleaning, refuse collection and public transport.
- private service companies, especially in the areas of cleaning, catering, security, refuse disposal, etc. who see their market shrinking because of the downturn in economic activity in the private sector. They see the public sector as a potential area of growth.
- capital goods manufacturers, who see the mechanisation of parts of the service sectors as possible markets for their new computer-based and high technology machinery.
- the users of services who see the often unresponsive and ineffective services provided by the state worsen as cutbacks and spending limitations reduce the adequacy and level of provision of collective services.

The outcome of the conflicts between these competing forces is obviously unclear. One thing *is* clear: the level of state expenditure on the services provided by the state is unlikely to maintain the rate of increase of the past decade over the next one. What sort of changes might be expected in the functions at present provided by the state?

An important dramatic change in recent years, particularly in Britain, has been the significant entry of private capital into the area of the economy previously accounted for by the state provision of collective direct services. For example, the private health care business in Britain accounted for £330m in 1981. Large service companies operate at an international level in the fields of cleaning and catering: for example, operating huge service-contracts in the Middle East and Mexico supplying street-cleaning, refuse or large-scale catering. Furthermore it has been government policy in Britain deliberately to force the introduction of private firms into state services in an attempt to reduce expenditure on them. Thus recent government directives have been issued to force health authorities in the National Health Service to award contracts for ancillary services such as cleaning, laundry and catering. At present private firms are competing against the public provision of these services at the expense of the wages and conditions of their workers and the quality of services provided. This has been found to be the case especially in cleaning. Commercial catering is becoming more capital-intensive with the introduction of 'cook and chill' and 'cook and freeze' systems of meal-preparation. This form of catering has applications in industrial and public catering and is also being considered as a replacement for school meals, as well as replacing meals on wheels (provision of meals in a state-funded voluntary service for the elderly and housebound).

The entry of capital into new sectors and the subsequent increase in capital-intensity is a well-established process within manufacturing but collective services are now experiencing changes along these lines too. The use of television monitors and electronic surveillance techniques in security work and the use of computers to optimise the movements of vehicles in refuse collection or public transport are further examples. It is not necessary for privatisation of services to take place before such transformations can occur. Capital-goods suppliers often have an important role in diffusing new techniques and technologies. (For example, British Oxygen Company, suppliers of liquid nitrogen, have strongly supported the development and use of 'cook and freeze' technology.) And these state services tend to have certain features which are ideal for technical and organisational changes, such as guaranteed, large and continuous markets for their services, attributes particularly appealing to private contractors, whose access to finance is greater.

RESTRUCTURING HEALTH CARE SERVICES

Introduction

We do not intend to explore all the possible scenarios for the transformation of state sector services here.[14] In the remainder of this chapter we will use one sector as a 'case study'. The role of information technologies in health care will be a useful example for some speculative discussion of how service provision might change in the future.

We should emphasise that our discussion *is* speculative. The technological developments we describe are not necessarily very well advanced in the sense that they are likely to diffuse rapidly throughout the health-care sector of any advanced capitalist country. Such diffusion would anyway require substantial changes in the organisation of the institutions which currently provide health care (such as the National Health Service – NHS – in Britain) in the attitudes of the medical professions to certain medical technologies as well as substantial public investment in the appropriate telecommunication infrastructures. These changes will not come about because developments in information technology 'demand' them; they clearly involve complex professional and political struggles. However, we seek to describe some of the strategic 'options' for the development of health-care services in advanced capitalist countries (although we focus principally on Britain) as seen from the perspective we have developed in this chapter – namely, the restructuring of services taking advantage of the economies offered by information technologies in the context of solving the long-term structural economic crisis of capitalist countries.

Over the past ten years, governments in all advanced capitalist countries have sought to slow, or even reverse, the growth of welfare expenditure by a variety of means. They include: the closure of inefficient facilities (rationalisation); an increase in the efficiency of ancillary services (such as laundry and catering) either by better management methods and the use of less labour-intensive equipment or by subcontracting these services to more efficient outside agencies (called 'privatisation' in Britain) and reforms in the management and financial structure of the services so as to introduce stricter cost-accounting methods.

The increasing cost of medical care has proved a problem for both the state-provided service and the private, profit-making sector; in the latter, increasing competition between private companies providing health insurance which funds the majority of private health care has put pressure on private hospitals to economise. The effect of these competitive pressures has been varied. One major factor differentiating private from public provision has been the quality of cost-accounting information available in the former. This allows high-cost procedures, both medical and administrative, to be identified more easily for cost-saving. As a result, even private hospitals are beginning to realise the dominant role that senior hospital doctors play in determining the costs of medical treatment (*Financial Times,* survey, 24 January, 1984).

In health care, the organisational changes that are under way also have a technological dimension, in both the administration and in the delivery of care. In administration for example, computers are being used in resource management to facilitate better planning. They are also used in keeping records of patients and in scheduling appointments. Such developments are extensions into medical administration of computer hardware and software applications already in routine use in other spheres.[15] But it is in the delivery of health care that the most interesting and socially significant technological innovations are taking place.

Much attention has been directed towards reducing costs by changing methods of surgery, the most profitable area of medical care in the private sector. This can be done either by the introduction of capital-intensive high technology medical equipment or by a expansion of day-care facilities, reducing the overheads of twenty-four-hour nursing care (currently around £100–£200 per night). New techniques which simplify surgical routines, such as the use of lasers, or even remove the need for surgery altogether, as in the use of the lithotripter to remove kidney stones, can be used to cut both surgery and after-care costs.[16] Also the profit-making sector has a role in trying out new technologies and care techniques, even though the use of such techniques in private medicine may be for marketing reasons, to give a 'high-tech' image, rather than because it has proved efficacious (see Thunhurst, 1982, pp. 43–6).

The process of transformation of the state-supported health care sector by private medical capital is likely to be a difficult business. As the critics of the growth of private medicine alongside the NHS in Britain have noted, much of the growth depends on the satisfaction of lucrative markets such as acute surgery at the expense of the NHS, in terms of the use of its trained staff and facilities (Iliffe, 1983). At present, resistance to the development of the private medical business within the public sector has led to the growth of the private sector outside it. Hence the massive burst of private hospital building in Britain over the last few years, which has doubled the number of private beds. Current attempts to reduce National Health Service costs revolve around the privatisation of ancillary services but this would still leave much of the medical core intact. To transform that would involve a major reform of the professional powers of doctors and, ultimately, consultants.

One proposal which has been mooted for this purpose is the replacement of the current managerial committees involving administrators, consultants and nursing officers in the NHS by professional managers. Such a radical proposal is sure to meet with resistance from senior doctors as well as from those workers in the NHS who favour more democratic forms of organisation. However, in private hospitals, such a management structure is more likely to be implemented particularly in the profit-making hospitals. In fact, current state policies in Britain seem to be directed at dissolving the barrier between private and public provision rather than dismantling the public service as such. So the outcome of these developments and similar ones in other countries is open. In some countries we may see a two-tier system of health care emerging, much like that in the USA, with a large private sector backed up by a rundown public sector acting as a welfare 'safety net' for the poor. Alternatively, there may be a restructured universal service, although perhaps provided in large part by private medical firms under overall state regulation. A more preferable outcome might be a better-funded and more responsive public service under wider democratic control.

Information Technologies in Health Care

Apart from the technologies of in-hospital patient treatment, there are technological developments based on information technologies which are of interest for the longer-term restructuring of health services: computer-aided diagnosis, automatic analysis and remote monitoring of patients. We elaborate on these in turn.

Computer-aided Diagnostic Systems

Computer-aided systems of diagnosis are already used in a variety of contexts in the formal health care system.[17] Such systems can be used as computer-aided learning techniques for training medical personnel. In health care itself they can be used to collect information from patients in advance of any professional diagnosis in a hospital or in a doctor's surgery or to assist doctors by calculating the relative probability of various diseases from declared symptoms. Computer-aided diagnosis has been shown to work in hospital environments at levels of comparable accuracy to diagnosis performed by doctors. There are also savings to be made by reducing the number of operations and tests which are otherwise performed unnecessarily. There are economic incentives which could stimulate the wider introduction of computer-aided diagnosis into the formal health care system. Diagnosis could be done more quickly, thus increasing the productivity of doctors or alternatively, diagnoses could be made more simply and could be performed by lower salaried para-medics.

However, the speed with which such systems are introduced will depend on institutional factors, such as the attitude of the medical professionals and on the technical constraints of the kinds of computer programs that are developed. There are two general methods of computerised diagnosis. One – the 'statistical' method – needs a certain level of clinical knowledge on the part of the doctor to interpret the computer result. In contrast, the 'logical' method requires the patient to provide information in answer to video-screen questions, as the program runs. The computer then produces its own result. So in principle, such programs could be run by GPs, para-medics and even by patients themselves rather than only by specialist medical consultants. The program would then be using the model of the doctor's analytical techniques, in a so-called expert system, which could be used by non-expert diagnosticians.

Automatic Analysis Equipment

Much of this equipment is expensive technology for use in hospitals. Examples are ECG, ultrasound and computer tomography and automatic tissue-fluid analysers. But some equipment for use outside the hospital or surgery is also being developed. Electronic machines for the measurement of blood pressure and pulse rate are already easily available. In the USA, ambulatory ECG monitoring for home use has been introduced with a market in 1982 of $85m. Sciencare Corporation launched a pilot compact ECG monitor in March 1982. This device is worn by the patient twenty-four hours a day; it measures blood presure and electrocardiogram signals which are stored on a cassette tape to be replayed by a doctor at her or his surgery on a specialised scanner. Although each unit is sold for $2000, the cost could drop as the market grows so that it can be used by a larger number of patients. Substantial progress has been made in home blood-glucose monitoring for diabetics. Blood glucose-monitoring machines are micro-electronic devices which use a photo-electric method of measuring the reflectance of reagent strips before and after staining with a small amount of blood. These tests can be done by the diabetic at home or at work and replace or supplement urine-testing which is less accurate and less convenient. Since the introduction of the innovation in 1978 in the UK, doctors and patients have realised the value of controlling the blood-sugar level helping to prevent blindness and renal failure which sometimes affect diabetics.

At present the British government is involved in the development of home blood-glucose monitoring machines. The Department of Trade and Industry has supported the firm Hypoguard in its development of a blood-glucose monitor, 'Hypocount'. Hypoguard also manufactures reagent strips for blood-glucose testing and use with the monitor. The company is now using its expertise gained through the development of Hypocount to manufacture micro-electronic counters for blind and partially-sighted diabetics. 'Dia-data' is a unit into which the diabetic enters up to thirty days of readings from the Hypocount, food intake and activity levels. The patient then takes her or his unit to the doctor whose master unit prints out the information with a summary of the patient's condition.[18]

Remote Monitoring

For patients with some chronic illness or longer-term condition (like pregnancy) frequent visits to the doctor or hospital can be inconvenient. This is true particularly if the distance from home to hospital is large as it will be in countries with scattered rural populations (Australia, for example). However, using various electronic devices, many conditions can be monitored at the patient's home, by the patient, and can be transmitted for hospital analysis via established telecommunications systems. We will give two examples from the UK. ECG monitoring of adults with dangerous heart conditions can be done 'at a distance'. Patients, on suffering chest pains, can telephone the hospital and, using an ECG monitor similar to the one already mentioned, can transmit its results down the telephone to the hospital monitor. The doctor can establish whether any heart attack is imminent and dispatch an ambulance if it is. A similar experimental system is in use to monitor the foetal heart in pregnant women. By means of an electronic sonic detector, the woman picks up the foetal heartbeat and transmits it live via telephone or, it is planned, by radio to a hospital computer which analyses the signal to identify any potential malfunction of the foetal heart. The whole procedure takes thirty minutes. The cost advantages of such remote monitoring techniques seem large – even including the cost of telephone calls and the use of computer equipment, each home monitoring costs less than 6 per cent of the daily cost of a hospital bed (which many of these patients would otherwise have to occupy to be under direct hospital surveillance).[19]

Clearly such developments, insofar as they increase the productivity of the health services, are likely to receive the support of health-care authorities who are trying to cope with slowly growing budgets. As they imply an increase in the capital-intensity of some aspects of that care they will also have some effects on employment and skill requirements in the health services. But they also offer new markets and, wherever the production sites are located, new jobs in the medical equipment, computer hardware and, particularly, software industries. The medical equipment industry is an expanding part of manufacturing. In Britain, the health-care products industry, which includes pharmaceuticals, is growing in real terms by around 7 per cent annually, though certain high technology sub-sectors are expected to grow at a much faster rate than this overall figure. For example, it has been estimated that diagnostic imaging equipment production will have expanded by nearly 180 per cent between 1975 and 1986.[20]

Will the widespread diffusion and use of information technologies of all kinds alter not just the way existing health services 'deliver' their care but also the *kind* of care that will be demanded by the population? Health care as a necessary human need can be satisfied within limits, either by the formal health *services* (like the NHS in the UK) or by other informal systems which tend to rely on privately-produced marketed services or physical products. In this informal system at present there is an increasing emphasis on preventative health care and on 'unorthodox', 'alternative' forms of medical care. Encouraged partly by state campaigns of health education and by the actions of pressure groups campaigning against forms of environmental pollution, there has been an upsurge over the past ten years in participation in various illness-prevention pursuits, such as jogging, gymnasia, sports centres, and increasing interest in lifestyles and eating habits appropriate to good health (more fibre eaten and less smoking amongst men!). While some preventative health measures are costless in that they involve a switch of spending between foodstuffs, for example, many of them involve the creation of new consumption tastes and the growth of some industries with technological change in others. For example, jogging stimulates the sportswear and equipment industries; increased fibre consumption switches breadmaking to wholemeal baking technologies; all of them stimulate new branches of publishing. In addition in the 1970s there has been a big increase in the number of people seeking health care advice outside the formal system. In the UK, the number of 'approved' practitioners of such techniques as osteopathy, herbalism, acupuncture, bio-feedback and faith healing has doubled over the past five years, to nearly 8000.[21]

The expansion of health services along preventative lines might be the only real way both to increase good health and reduce the cost of medical expenditure in the long run, as Doyal (1979) has pointed out. But although the importance of preventative aspects of medicine in Britain has been recognised by the state in terms of policy – though with little funding – it is the private sector which has moved into this vacuum with most speed.

There are two possible scenarios conceivable in the development of these two health-care systems, the formal welfare service and informal preventative and private advice sectors. One assumes that the balance will remain much as it is now. There will be substantial productivity improvements in the formal sector which, although still a relatively labour-intensive service industry, whether state- or privately-financed will continue to provide for most people's medical needs. (Some of course will still seek unorthodox advice and others (many) will continue to medicate themselves and participate in self-help activities.) The other assumes that whatever may happen in the welfare-service side of health care, the trends to *self-service,* to purchase equipment and information to care for one's own health, will expand considerably, but will also be simultaneously restructured around information technologies.

Over the next twenty years therefore there could be expanding markets for high technology health care products to be sold, not just to or through the formal health services – though this will be large market – but also on a mass-consumption scale to *individual consumers* directly. The development of these markets, implying the manufacture and maintenance of electronic equipment and the production of all kinds of software to run on it, offers the prospects of new employment at the same time as employment in the formal health service may not be growing, although whether such markets will be mass markets is an open question.

The balance actually struck between the two modes of provision of health-care will depend on many factors; a crucial economic determinant however will be the balance of cost between the various modes. There are currently a number of options available to the would-be patient seeking, for example, diagnosis of some complaint:

1. to seek the advice of a medical practitioner, either in the orthodox or unorthodox sectors;
2. to seek free advice from friends, the pharmacist etc., and if necessary, purchase proprietary drugs;
3. consult medical books and proceed as in 2).

(1), (2) and (3) are in no way new options, the choice between them depending on a variety of factors, including the costs of consultation and of the alternative treatments. Clearly developments in medical equipment might reduce these costs or at least stop them growing. The next two, certain technological developments permitting, *are* new:

4. consult bought or rented computer programs and/or perform routine tissue-fluid analysis/blood pressure (etc.) measurement and self-diagnose; and follow therapy outlined. (This could, of course be some unorthodox therapy recommended by the computer program which was specifically purchased with this in mind; or be a recommendation to consult some medical practitioner.)
5. as (4) but the whole activity carried out at some self-diagnosis centre not necessarily connected with the formal health care system, but possibly provided by private health-check companies.

Clearly (4) and (5) are not economic options at the present time.[22] They rely upon the generalised possession and active use of, or access to, personal computers and on the provision of substantial information technology infrastructures. Although these are being installed and the diffusion of personal computers is becoming more rapid, it remains to be seen whether they could ever be used routinely for anything as sophisticated as the provision of health-care information. A critical technological factor would seem to be the availability of routine techniques of self-diagnosis. As described above, such techniques are currently being developed, although mostly for use in the formal health-care system. Nevertheless there are some signs that larger markets are being sought. Testing and self-monitoring equipment of various sorts is available – particularly for accurate measurement of blood pressure, pulse rate and some body-fluid analyses as well; but it is not yet particularly cheap. The link between hospital computer-aided diagnosis and home diagnosis has not yet been commercially developed but during the next five years we can expect a growth both in the number and range of such systems, particularly as some of the conceptual problems are attacked by increasing research into 'artificial intelligence' (see Alvey Committee, 1982; Feigenbaum and McCorduck, 1984).

There are already a number of computerised diagnostic programs available to those with video-disc players and home computers. The pharmaceutical company Smith, Kline and French has produced an interactive video-disc to educate doctors in the field of gastro-enterology (The disc however is extremely expensive at £28 000). A number of computer software firms offer for sale computer programs designed to aid doctors in the diagnosis of various psychiatric conditions. (They are known as 'HEADACHE' and 'FREUD'!). There are also a number of programs available to home-computer owners such as Network Computer Systems' 'First Aid Program' and Eastmead's 100 program 'Home Doctor' series which, claims Eastmead, 'aims to complement the services offered by the NHS by providing patients with a series of simple health programs which differentiate between those symptoms requiring medical attention, emergency and routine, and those which could safely be treated at home without help (*Medical News*, 7 April, 1983).[23]

Self-diagnosing computer programs are thus available but they are not very sophisticated and are certainly not able to recommend very detailed courses of treatment. Nevertheless these various techno-logical developments make the self-service option within health care more possible, subject of course to other economic and political decisions providing both the infrastructure and the incentives for their development as mass market commodities. Of course, it is possible to envisage various *intermediate* stages in any path of development to self-service diagnosis, etc. – the high cost, in the formal health-care system, of continued analysis for some conditions may make self-analysis technologies an economic option within a more cost-conscious health service. Computer-aided diagnostic systems may be best operated by para-medics, screening out patients with routine ailments but under the professional scrutiny of more highly-trained doctors possibly within a private health care system. Vital here will be the attitudes of the medical profession towards what is quite clearly a challenge to their monopoly over medical knowledge (Child *et al.*, 1983). Whatever they do, it is quite conceivable that preventative health programs, of both the computer and video kind, compiled with keep-fit routines and information on healthier lifestyles could infiltrate the home via the computer, video and cable TV markets.

Whatever are the various paths whereby such equipment will come into routine use outside the existing formal health-care system, it does seem clear that information technologies will change in some fashion the way the population seeks to satisfy its requirements for health care. The questions remain as how, to what extent and for which social groups? Of course, any increase in the informal, self-service sector will depend on the relative costs of the two sectors and this is linked to the policies of the health-care authorities, of the health insurance companies, and of the medical profession.

CONCLUSION

We have tried to show in this chapter how new electronic and telecommunications devices, with the appropriate software of course, can have profound implications for the way in which service production is organised over the next few decades. Indeed, far from the post-industrial scenario of a highly mechanised manufacturing sector supporting a huge, low productivity, labour-intensive sector of service work, information technologies provide one of the means in which the service sector itself can show large rises in productivity. The service sector is the site of major upheavals in the forms of organisation of production in advanced capitalist societies.

In clerical work, self-servicing externalisation and the welfare services, there are two aspects to these upheavals: first, the nature and structure of the services provided, in particular the changing mix of waged service labour, machinery and unpaid self-servicing labour, and second, the labour processes whereby the waged service labour delivers the service product. Information technologies do not of themselves determine the outcome of the struggles, between firms or between workers and managements, over these incipient changes, but they do change the range of options for profit-making in services and thus present new areas and issues over which the struggles can take place. So by making possible the more rapid processing of variable information by means of advanced computers and communication networks, types of service product which required relatively large numbers of experienced information-handling workers can be offered more cheaply.

However, such productivity increases need not be gained by

Fordist organisational methods of job simplification and standardisation within some hierarchical management system. Much more diverse forms of production and work organisation become possible, although we need not assume that the favourite of progressive management consultants – autonomous work-groups integrated by interactive computer networks – is the only one. As we have described, organisationally highly centralised but geographically dispersed forms of production organisation – such as 'tele-outwork' – despite their questionable advantages for the outworkers are equally possible, along with 'satellite' offices, for some types of clerical work at least. It is our contention that whatever organisational forms might be adopted to exploit the potential of information technologies, they will be best understood as marking a discontinuity with Fordist forms of organisation that have previously been devised for some service industries; hence our emphasis on neo-Fordism.

But such reorganisation of service labour processes will imply changes in how the service product is delivered and indeed the nature of the service product itself. The externalisation of service labour, with individual consumers providing at least part of the service themselves, is already established in retailing and mass catering and is likely to expand rapidly into the information services as higher capacity telecommunication infrastructures are installed and domestic terminals and input devices fall in price.

As we have tried to show in our admittedly rather speculative discussion of possible information technology-driven changes in medical diagnosis and health care, information technologies in their application are not just limited to improving the labour productivity of welfare services as presently delivered. Much more substantial changes, satisfying human service functions in radically new ways, are conceivable, although, as we have pointed out, such changes will not inevitably occur just because they can be imagined technologically. Nevertheless it is clear that as new profit opportunities are sought in the current depression, the application of information technologies to service industries is a prime candidate; changes in the modes of service delivery – the switch from labour-intensive service provision to new mixes of computer-based equipment with intermediate services and self-service labour – are at the centre of making information technologies profitable. Any upswing for capitalist economies must be based not just on some restructuring of existing industries but on finding *new industries* in which to make profits.

8 Conclusion: From Fordism to Neo-Fordism?

In this book we try to show and investigate seven things:

1. We demonstrate how technological developments of the past hundred years can be seen as a progression of overlapping phases of primary, secondary and tertiary mechanisation, exemplified in particular leading industrial sectors; those developments seen within certain work organisational contexts, in labour processes, can be shown to link with long waves in capitalist economic growth.
2. We argue that the technological basis of the post-World War Two boom period of the fourth long wave can be seen as the generalisation of secondary mechanisation with the emergence, in a small number of industries, of tertiary mechanisation.
3. We identify Fordism as the dominant labour/production process paradigm of the boom period of the fourth long wave.
4. We see the depression of the 1980s as one of a crisis of a specific form of capital accumulation, based on particular Fordist production methods, products and consumption patterns which came from and led developed economies out of the depression of the 1930s.
5. We examine the limits of Fordism and how these might be overcome using new technologies and emerging new forms of labour process organisation – the production process paradigm we characterise as neo-Fordism.
6. We look at emerging forms of neo-Fordism in small-batch production and in the service sector and how new technologies offer new ways to organise both production processes themselves and, most significantly, the products that are available for mass consumption.
7. We have argued that any resolution of the present crisis which maintains the viability of capitalist economies, restoring capital

193

accumulation and producing a new upswing in the economy must involve further radical changes in production methods and in consumption patterns over the next decade.

In pursuing these ideas we have sought to bring together two areas of thought – labour-process analysis and long-wave theory – to show, we hope, how an adequate understanding of the current economic crisis demands some comprehension and integration of the insights of both areas. So we would claim it is not possible to understand the nature of the current depression (or, if you like, the downswing in the current long wave) without noting the importance of changes in production processes, seen as combinations of specific technologies and specific forms of the organisation of work (and therefore of workers). Further, the 'trigger factors' which might stimulate any upturn cannot be read off from the pure technological-innovation options identified in the depression. Rather we have tried to show that, at least in previous lower turning-points, the crucial enabling factor in any upswing was a specific combination of those technological options with appropriate forms of labour organisation which offered better profit opportunities for expectant investors. Since labour organisation is inherently political, involving such things as the balance of power between capital and labour, the types of labour required and available, the regional distribution of industry – all matters of local, regional, national and international political concern – then political factors are as vital to the upswing as technological options. We see the political programmes being pursued by right-wing governments in OECD countries, particularly in Britain and the USA, at the present time in this way – as attempts to reduce the power and wages of organised labour by specific laws and actions against trade unions, by the discipline of large-scale unemployment and by the recasting of educational and welfare programmes to lower the subsistence levels of large numbers of the population. Of course, all such political events have unpredictable outcomes. Indeed, as we will discuss later, other possibilities which do not accept the priorities in the exploitation of new technology's potential set by capitalist firms and their supporting governments, are conceivable.

But nevertheless – and this we definitely insist on in this book – the technological options are not merely subordinate to these political or labour organisational factors. Scientific knowledge, and any technological innovations which might develop from it (although

not necessarily in the format: 'science discovers, technology applies')
is not completely random and unplanned; capitalist firms, with
government support, do their best to focus scientific and technological research and development on to soluble problems of product-
innovation and productivity-increasing process design and infrastructural provision. But there is inevitably some indeterminacy in
their planning attempts, for all sorts of reasons – not enough money,
bad planning, choice of wrong problem, inadequacy of scientific
knowledge, insufficient trained people. Only some research programmes are successful, only some technological options are finally
presented in an economically viable form as successful innovations.
In short, technological options available to capitalist firms at a
particular time are bounded, and although their choice and exploitation cannot 'guarantee' (and retrospectively cannot be seen as an
'explanation' of) any upswing, the other determining factors within
the labour process cannot guarantee or explain it either. A study of
the *technological* options which may be involved in a solution of the
problems of the current depression can surely illuminate what
broader *political–economic* options are open, so long as these options
are seen as diffracted through changing production/labour processes.
If seeing the importance of technological options makes us 'technological determinists', we accept the appellation.

In discussing labour-process analyses of production changes this
century we have argued for some broader dynamic framework for
these changes. Braverman, from whom most labour-process analysis
springs either to illustrate his general claims or to refine him by
partial refutation, saw his discussion of the causes of the 'Degradation of Work in the Twentieth Century' (the sub-title of his book) as
part of the broader process of the monopolisation of capital since the
nineteenth century. Many subsequent studies of individual cases of
the changes in skilling levels consequent on industrial restructuring or
technological change have not brought in this politico–economic
dimension. Or if they have, it has been subordinate to perceived
underlying trends in the capitalist economy which have continued to
seek to divide conception and execution, to deskill previously skilled
craft workers.

We have challenged this idea of technological change having a
deskilling dynamic under the constraints of accumulation and
competition, in two ways. First, we have focused attention not on the
labour-process analysts' favourite of *Taylorism*, a mode of labour
organisation and task allocation, but on *Fordism* which implies

certain forms of labour organisation and task allocation but within a broader framework which brings together the nature of the product, the market size and, linking with this, the forms of consumption of the wider working class as well. Second, in Chapter 4, we have tried to emphasise some *periodicity* in the development of capitalist labour processes over the past 150 years by identifying various phases in the trajectory of the mechanisation element of the labour process (which, we must repeat, we do not see as determined, in the crude unilinear sense, by machine developments); and we have attempted to link these mechanisation phases with even broader politico–economic trends as exhibited in the upswings and downswings of long waves.

We would not like to claim too much as the consequences of our analysis. By definition, by identifying the long-wave lower turning-point as historically contingent, much more so than the upper turning-point, no general projections of future developments can easily be made. We would not therefore feel secure in proposing specific policies to 'bring forward the upswing', though others have (Mensch, 1979, for example). However, we would claim that from an understanding of the trajectories of mechanisation and of the Fordist underpinnings of the past fifty years some general issues emerge for consideration by those like us who are concerned with political transformations within capitalist countries.

Chapter 7 suggested that it is reasonable to ask whether the current economic crisis, the recession–depression phases of the long wave as some would say, is associated with some 'exhaustion' of the Fordist paradigm established with the 1950s' boom. Have Fordist production processes reached some limit both technically and in terms of labour management? Are the forms of organisation and the patterns of consumption that constitute Fordism 'appropriate' for the optimally efficient use of the new information technologies? More particularly, is Fordism, the historical form of production organisation which corresponds to a certain stage in the development of mechanisation and a specific configuration of labour organisation, 'suitable' in the sense that it is appropriate for renewed accumulation, if information and other technologies are to be applied successfully to those sectors of production which have hitherto not benefited from it?

There is a wide literature, dating from the late 1960s as we have described in Chapter 5, which suggests that Fordism (or Taylorism) has had its day. Work redesigners, job enrichers, socio-technical theorists have all argued that production processes based on 'automation', in firms which need to adapt more quickly to their competitive environment, cannot depend upon techniques of management organisation (mechanistic Fordism–Taylorism) devised in very different technological and competitive circumstances. However, few have linked the crisis in Fordist systems to any overarching economic crisis or any phase of longer-term economic cycles, preferring mainly to argue that the new production organisations are more humane, more mentally healthy, more democratic or more in tune with an educated, affluent, modern workforce.

Gordon, Edwards and Reich, however, have indeed tried to do this in their examination of the history of the American working-class, or more particularly of its divisions. They examined the interaction between what they call 'long swings' in economic activity, the organisation of work, labour market structure and what they call 'social structures of accumulation'. By the latter they mean the 'specific institutional environment' within which capital accumulation takes place. Long swings and social structures of accumulation are connected.[1] As they state:

> We shall propose that long swing and social structures of accumulation are interdependent and mutually determining in capitalist economies. A long period of prosperity is generated by a set of institutions that provides a stable and favourable context for capitalists. This context must provide capitalists with both profitable investment opportunities and a stable societal environment in which to realise them. The boom begins to fade when the profitable opportunities inherent within the existing social structure of accumulation begin to dry up . . . Long swings are in a part a product of the success or failure of the social structure of accumulation in facilitating capitalist accumulation. (Gordon, Edwards and Reich, 1982, p. 9–10)

The concept of social structure of accumulation needs further elaboration. Gordon *et al.* argued that it comprises three types of institutions. First, there are the systems which guarantee the supply and circulation of money and credit, the banking and finance systems. Second, there is the pattern of state involvement, whereby state institutions, either through tax allowances and industrial subsidies, procurement policies, training programmes and regulations assist in profit-making or, through laws and regulations restricting behaviour considered socially or politically unacceptable, reduce profit-making. Third, there is the political system through which class struggles – in unions, political parties, riots, pressure groups and elections – express themselves. They are of the view that what they call the 'post-war system of bureaucratic control' has slowed down US labour productivity and is thus involved in the current economic slump. Major structural changes in the American economy are necessary to produce a new social structure of accumulation which, American capitalists would hope, would enable some overall increase in social productivity and profitability and set accumulation on a new upward trend – in short to provide a new sustained upswing in the American economy, a new boom.

Gordon *et al.* are only able to give the briefest indications of what US capitalists are experimenting with in institutional and production organisation change which might, in some combination yet to be fought out in political struggles in the USA, emerge as an appropriate social structure of accumulation. At the national level, they suggest that current Democratic Party flirtation with ideas like incomes policies and a semi-planned industrial policy might in some corporatist compromise be an essential element of any political settlement (Rohatyn, 1984). At the corporation level, it is possible that various forms of worker participation in management might in the longer term be necessary to get workers in unions in established industries to accept drastic restructuring; also, the growing interest in 'Japanese' forms of management might presage some adoption of such methods, both in terms of shop-floor organisation and in the growth of labour-disciplining sub-contracting. Freeman has similarly suggested that for any economic upturn a large number of institutional changes are required, involving changes in:

> the pattern of skills of the workforce and therefore in the education and training systems; in management and labour attitudes; the pattern of industrial relations and worker participation; in working arrangements; in the pattern of consumer demand; in the conceptual framework of economists, accountants and governments, and in social, political and legislative priorities. (Freeman, 1983b)

Gordon *et al.* also argue that any new social structure of accumulation would be 'deeply affected by recent developments in technology'. They claim:

> Although it is still difficult to discern the implications for the social

relations of production, it is clear that recent technical innovations such as robotisation and the proliferation of micro-electronic equipment, carry profound potential implications for the organisation of work and the relative power of managers and workers. (Gordon *et al.*, 1982, p. 221)

We have tried to elaborate on these recent technological developments in this book but we have said little about any emergent social structure of accumulation, either in the USA, Britain or at the general international level. The concept of Fordism we have presented, whilst clearly implying an appropriate social structure of accumulation – so as to provide necessary credit for large-scale purchases of mass produced goods or to provide appropriate infrastructures and welfare programmes for example – is not intended to specify all the necessary institutions in a social structure of accumulation. Instead we have chosen to keep Fordism to a description of a process of labour organisation and associated consumption patterns.

If the last boom can be characterised as being structured on the combination of a specific technological form (the phase of the maturation of secondary mechanisation and the emergence of control mechanisation) and a specific system of labour organisation, all within the Fordist production process, on what might any new boom be structured? As Chapters 5, 6 and 7 argue, we see neo-Fordism within production processes as the likely combination. Neo-Fordism has three elements: the technological element of control mechanisation which permits the mechanisation, at high levels of productivity, of more flexible production of a higher variety of products; the labour-organisational element confers greater choice in the combination of tasks into jobs and work-roles thus moving away from individual repetitive jobs; the informational infrastructure element permits the integration of different productive sub-units by electronic methods. In addition, we have argued for seeing changing patterns of consumption, particularly of service industries, as an aspect of neo-Fordism.

As Aglietta has pointed out, the current economic crisis is partly a result of the 'crisis' of Fordism comprising both the resistance of workers to Fordist work organisation and barriers to the extension of Fordism into small-batch capital-goods production and the production of labour-intensive services, particularly those collectively consumed. Given the extent of this crisis, the process of the restructuring of capitalist economies for a new upswing would have to overcome these complex and interrelated barriers to accumulation. We have discussed these in Chapters 6 and 7 in relation to possible neo-Fordist changes. However, neo-Fordist developments do not rule out the further extension of Fordist systems to production processes in newly industrialised manufacturing sectors in developing countries – what Lipietz (1982) has called 'global Fordism'. He argues that since Fordist production processes separate out the unskilled assembly of products from the skilled production of parts and the conceptual tasks of design and planning, then in pursuit of lower costs these different parts of any process can be located in different countries. Firms in mass-production industries have thus sought out sources of cheap labour in the so-called 'newly industrialising countries' (NICs) and have applied Fordist and Taylorist techniques to labour organisation there. Whether this will continue in the NICs and spread to other third world countries critically depends on the struggles of workers in those countries to raise wages and resist the worst features of Fordist mass-production organisation (what Lipietz calls 'bloody Taylorisation'). But it also depends on political and economic developments in the advanced countries. Neo-Fordist changes, apart from offering markets for new manufactured goods and intermediate consumer services, are associated with the reorganisation of those industrial sectors not yet transformed along Fordist lines. It is possible to envisage neo-Fordism in the advanced capitalist countries coexisting with Fordism in the NICs, though we do not insist on the division.

To conclude we shall now summarise our discussions drawing out some broader conclusions indicating the issues our analysis would suggest will be on the political agenda over the next ten years. We refer to issues concerning the response of workers, particularly 'craft' workers in both the conceptual and executive aspects of production, to the introduction of new technology into their workplaces and concerning changes in service provision in the public sector that consumption restructuring signals.

WORKERS' STRUGGLES AND THE PROBLEM OF 'CRAFT'

As we have already pointed out, product changes and productivity increases in capital goods production have 'ripple' effects through all sectors of the economy because of the lowered cost of machinery or the increased productiveness of new capital goods produced. Such a process is a continuous one but in a depression, as firms rationalise and as profit expectations change, there is an opportunity for a wholesale re-equipping of production processes, using machinery based on newer technologies. This point is crucial to a characterisation of technological change during a depression period leading to an upswing and forms the basis for our stress on developments in mechanisation. As we demonstrated the potential increases in productiveness which particular new technologies bring about are not achievable without wholesale changes in the products, organisation and techniques of production within particular branches of industry. The scale and problems of such changes and thus the actual rate of change have not been appreciated by many writers describing the likely timescale of the micro-electronic revolution who assume the widespread use of 'bolt-on' labour-displacing control systems in manufacturing industry and the computerisation of white-collar functions leading to 'automated' factories and massive unemployment.

Periods of restructuring, then, result in the emergence of new industries but also involve related shifts in techniques of production in many existing industries. As we have shown, this is the case with small-batch engineering where this 'backward' sector (in that it did not go through a Fordist phase of development) is being 'leapfrogged' into a neo-Fordist phase by the use of computerised technologies. A similar case might be made for printing in the British national newspaper industry as well as for many other so-called 'sunset' industries; neo-Fordist techniques could well allow them to rise again, rather than be effectively relocated to other countries, although levels of employment would still be lower than in those industries during the last boom.

Such neo-Fordist developments put in doubt the trajectory of 'deskilling' that might have been observed previously or might have been expected as Fordism 'crept' into those industries. Those who identify such Fordist tendencies, such as Kraft (1977) in computer software production, Cooley (1980) in engineering design work or Barker and Downing (1980) in clerical work, suggest that deskilling should be resisted and existing craft-based job-roles should be defended. However they do not necessarily reject the use of new technologies completely, suggesting instead 'capturing' strategies in which the new technologies are operated by existing workers under their control (a strategy which has been tried with some success by predominantly male typesetting workers in Fleet Street).

There is certainly much scope for the further development of these strategies if you assume, as we do, accepting many of the arguments of socio-technics writers discussed in Chapter 5, that the organisational choices in the use of new technologies that managements make are not always the most appropriate even for assured profit-making. The design of software to operate the technological infrastructure of neo-Fordist systems is one area where design criteria 'alternative' to those currently conceived by managements can be invoked. Some Computer-Aided Manufacturing (CAM) systems incorporate routines for carrying out synthetic timing of jobs ('built-in time and motion study') a political choice by management to embody in the system some means to control piecework payments. Yet, rather than CAM taking the form of a one-sided *management* information system which collects data on the work-performance of individuals for disciplinary purposes as well as monitoring and co-ordinating the flow of work through a machine shop, the former functions can be eliminated from a system in the process of bargaining over its introduction. Further, attempts by management to restrict access to the database, excluding shop-floor workers, are also negotiable. The relative technical ease of modifying informational systems opens up opportunities for the extension of collective bargaining to wider issues than pay and conditions. Compared with the difficulties of contesting the design and layout of production machinery where these had limited flexibility, there is a high degree of choice in design and operation of the new computerised machines and information systems.

However, to contest change of this substantial technical kind requires considerable expertise regarding the likely effects of the new systems being installed and possible alternative forms of utilising them; this expertise tends to be concentrated in technical workers who, while becoming more unionised in recent years, have generally failed to ally themselves to broader working-class movements and tend to have a technocratic view of the use of new technology. It is surely an important priority for trade unions in all countries to develop structures and procedures whereby the expertise in computer systems design and associated technological infrastructures can be obtained and workers in possession of that expertise be encouraged to consider alternative design criteria.[2]

Such strategies, to regulate the introduction of new technologies into existing industries and seeking alternative design criteria, can only be partly successful however, for these reasons. The severing of the Fordist link between mass-production and economies of scale, allowing the production at lower costs of smaller but still economic batches of both goods and, most significantly, services – which until now have depended on physical concentrations of workers in large production units – is one of the technological bases of the possible reconstitution of production on a new basis. The number of workers required may be reduced (per unit of output and thus, unless markets expand substantially, absolutely); the types of skills required may be different and may thus be found from different pools of labour-power in different locations.[3] But communication problems, the informational links between sub-units of an overall production system or between a system and its suppliers will be made easier by advanced electronic devices and telecommunications infrastructures.[4]

Thus 'deskilling' can occur not through direct confrontation of new technologies with craft workers but rather through industrial rationalisation and redundancy, in parallel with the introduction of new employment using different, cheaper, more amenable workers. Further, much new technology is likely to be introduced into the service sector, and it is traditionally in those sectors where union membership is low and, in particular, there is a low level of craft union tradition.

However, focusing on craft strategies in resisting and controlling new technology reveals their divisive nature. Strategies for controlling labour supply through, for example, apprenticeships and closed shops with restricted conditions of access, or for ensuring a supply of work through, for example, demarcation, have succeeded in the past but at the expense of less well-organised sections of labour (see Elger, 1979 and Cockburn, 1983). There are real differences of interests between sections of the working class, so it is necessary to devise new forms of democratic decision-making within which these differences can be resolved rather than concentrating on further developing *competitive* strategies.

In fact in view of the radical cross-sectional implications of new technologies, we would argue that it is often impossible to devise successful employee strategies at the level of occupational group or, often, even at the level of a particular firm or industry. One of the most innovative political developments of recent years has been the emergence of workers' plans within large firms in particular industries. This type of strategy emphasises wide grass roots involvement in the construction of alternative corporate plans. The plans incorporate alternative decision-making criteria in the design of products, organisation of production and R&D policy which are oriented to the provision for social need, democratic control of production and participation in choices associated with the process of research and development. Such ideas have been taken up in recent years by left-wing local authorities in many countries, a prime example being Greater London Council.[5]

There are of course a number of difficulties in the construction of such alternative corporate plans. Not least, is the problem that in many firms research and development as such is actually carried out in the firms of their capital goods suppliers, many of which may be foreign-based or foreign-owned. In Europe, as Holland (1983) and Radice (1984) have pointed out, this implies that radical policies which are seriously to affect the strategies of large corporations are pointless if restricted to the level of national states – a European dimension is more than ever necessary to control the introduction of new technologies.

SERVICES, JOBS AND WELFARE

Whereas production in the expanding industries of the post-World War Two era was very labour-intensive, the new micro-electronic-based goods of the 1980s will be produced in highly mechanised plants which are hardly likely to absorb the labour surplus from the job-shedding industries of the past ten years. Worse, as we have pointed out, mechanisation within service industries would seem to point to extensive job-loss in these industries which up to recently have been expanding in employment. Some have responded to this by arguing against the introduction of such job-destroying new technologies in the first place; others, accepting the 'inevitability' of technological change, suggest that the introduction of new technologies should be planned and ways found to cope with the large number of otherwise permanently unemployed by the better distribution of existing working time and by 'education for leisure'.

Gershuny and Miles (1983) have yet another view. They argue that with the installation of the telecommunication infrastructures currently under way (digital transmission systems, cable TV, satellites) many direct producer and consumer services will indeed experience rapid rises in productivity which will cut their costs and more than likely their employment. But there are considerable prospects for *new* employment in industries which supply the hardware that will plug in to the new telecommunications infrastructures, such as home computers. More intriguingly, new jobs might come from new service industries that will produce the software in the forms of TV programmes, computer programs and information packages – the so-called 'intermediate service products' – that consumers will use to satisfy those service functions which are at the moment still provided by direct service industries. So in this 'social innovation scenario', investment in a new, large-scale but flexible telecommunications infrastructure may not produce many jobs in the short run but in the longer run there would be many jobs in the production and distribution of software. Initially, the new domestic equipment installed in each household would be for entertainment and for further education. However, some services might grow as information about them became more accessible (for example, fringe arts, library reference); others might spring up later, such as computerised car-sharing pools, community shopping. (The list is not long.) If such constructive leisure facilities presented themselves, then reduced working-hours might be more acceptable, increasing the attractiveness of job-sharing and, acceptable wage-levels permitting, reduce unemployment somewhat. There would thus be an employment shift to the provision of intermediate consumer services.

Gershuny and Miles' scenario therefore sees service provision as being transformed from a different direction by the diffusion of information technology equipment into households to substitute for those services previously externally provided but now either cut back or too expensive. Services become software commodities purchased to run on mass-produced information technology devices. So, as new technologies profoundly change the means of producing physical goods and services, new consumption patterns are also established that imply a radically different mix of service-provision. The self-service trend provides new markets in the medium term for manufactured goods and new consumer services all of which, after some unknown transitional period and after some changes in the length of working day/week/year/life, forms the technological core of a new socio-technical system that is the basis of a new capitalist boom. At the same time a wide variety of services previously provided by the state – in education and health care for example – are, in a radically new technological form, brought into the household along with other information services, creating new large markets of individual consumer purchases for private firms, permitting a reduction in state provision yet further.

Now clearly this self-service scenario for a capitalist resurgence is not an unmixed blessing. As Gershuny and Miles point out there are crucial questions to be asked about distribution and access (what social groups will be the first and the largest beneficiaries of this change of service provision and who will be the losers? What regional disparities will be intensified? How will privacy be ensured?). One might add, how is the neo-Fordist self-service trend tied in with the trends to returning various caring services 'to the community' or towards more 'homework,' both of which hide the reality of the increased burdens such trends place on women – for self-service read 'housebound-women serviced'?

There are of course many open questions. For example, how should the information infrastructures upon which new products and services depend, be designed. In principle it is possible to envisage political campaigns through which social rather than market criteria can be brought to bear on what telecommunications options are selected. However, whereas one can keep an open mind regarding information technology's potential to assist such things as 'citizen initiatives to achieve self-managed health' and tele-information systems which provide social security claimants with better information on their rights, etc., it is not clear what one's attitude should be to the very idea of the restructuring of services along neo-Fordist lines. In the current political situation, as we have tried to show in our discussion of health care, it comes within the context of the cutting of the collectivised non-marketed services without it being obvious that better alternatives for the majority of the population are going to be on offer from the information technology sector. The rising cost of the public services and the bureaucratic administration of them which encourages people to seek 'private' alternatives, whether these be private direct services (such as, private medical care) or information technology-assisted self-services, pose political questions.

It is reasonable to argue that the trend to the self-service, 'commoditised' provision of service functions has in many cases gone too far already. In transport, for example, the disbenefits of mass individual automobile use have long been known – an increase in the provision of collectively-provided services by an *expansion* of metropolitan transport systems, perhaps using information technologies to facilitate booking and routeing, etc., might make more sense in terms of resource usage and urban planning (see Best and Connolly, 1982, p. 86–92). And as it is inevitably more labour-intensive than manufacturing goods and less skill-intensive than software production the prospect for job creation may be better.

The reconstruction of consumption in those services that were reorganised in the last boom is, however, uneven. In Europe, although as much as 80 per cent of expenditure on travel is accounted for by the purchase and running of motor cars, 20 per cent is still used to purchase final transport services. Part of these will be accounted for by non-domestic consumption which is undertaken by choice, for example, train travel rather than car travel for long distances. However, the rest is accounted for by the *failure* of provision by Fordist commodities of public transport for the poor, elderly or disabled – or members of a household who do not have privileged access to the family car. For these non-consumers the provision of collective services is crucial and thus the relative reduction in these services over the post-war boom is the other side of the Fordist expansion of markets.

For many services, their low productivity, reflecting low capitalisation and difficulties in transformation, has resulted in a significantly low rate of profit and limitations in the market growth of many of these final services. In an attempt to respond to the high costs of service provision and delivery, both private firms and the state have responded by tapping low-wage segments of the labour market. By using young workers, part-time labour (often married women with child-care and domestic responsibilities) or ethnic labour at a disadvantage in the labour market, many services have been populated with low-wage workers with little training in an attempt to increase profits in absolute terms.

This is particularly visible in low-skill service sectors such as cleaning, catering and retailing. But it also applies to the lower levels of professional services too. In state services, where labour organisation tends to be easier, it has been possible for workers to raise their wages and improve conditions a little, in comparison with other service occupations in retailing and private cleaning firms. Where profits are high in services, this is at the expense of the majority of service workers and the cost to users of services in terms of price and quality. It is clear that the necessary condition for this situation to be changed must be a way of breaking the blockage to productivity increases in service provision and delivery. Obviously, information-technology-driven mechanisation is one such way; but what is not certain is the effect of such changes on wage levels, if service workers are not better organised.

But to say this is not to suggest that information technologies have no role to play in providing better and more extensive public services. Campaigns to maintain or restore high levels of public provision and to introduce more democratic forms of management of the services, however vital, need some view on how new technologies can be utilised to improve the productivity of those services. Such productivity improvements can allow the increased wages of service workers, as well as providing the basis of an extension of public service in more labour-intensive functions. Successful application of information technologies to ensure some historic increase in productivity in the various sectors we have discussed (thus possibly contributing to some upswing in economic growth in capitalist economies) depends on the successful assembly of the elements of neo-Fordist configurations of production organisation. The exact forms of such configurations are a matter for speculation; in addition, it is not assured that such a configuration *will* be put together as there are technological and labour organisation problems. The labour problems, which vary from country to country, are concerned with such matters as the availability of labour at the right wage levels and with the necessary skills, and these in turn are linked to national and international economic and political developments. The resolution of these problems in favour of capital is not reducible to the technological or labour organisational level. In that they involve political and economic struggles it is possible that different political conceptions of production and industrial organisation could make their mark. What we have tried to show in this book is that although people make history it is useful to examine from which options, in this case centring on technological ones, they have to choose.

Notes and References

1 INTRODUCTION

1. For analysis of the re-evaluations of post-war technological developments, see Johnston and Gummett (1979) Part II.
2. See, for example, Evans (1979), Marsh (1981b), Friedrichs and Schaff (1982) and Stonier (1983); a comprehensive selection of readings on the 'micro-electronics revolution' can be found in Dertouzos and Moses (1979) and in Forester (1980).
3. 'The transition from formal to real subordination in the labour process' is a set of concepts borrowed from Marx's analysis of the development of the factory system; see Brighton Labour Process Group (1977).

2 ON NEW TECHNOLOGIES AND 'AUTOMATION'

1. See the Appendix to Blackburn, Green and Liff (1982).
2. Reviews of energy options for the next century can be found in Foley (1981) and Elkington (1984); the sea as a source of energy and materials is discussed in Ford *et al.* (1983); for a recent survey of developments in new materials see *New Scientist*, 26 January 1984, pp. 10–22.
3. See Yoxen (1983), European Community Commission (1983) and Zimmerman (1984) for British, European and American views of biotechnology developments.
4. Case studies of the changes in levels of employment resulting from the use of micro-electronics in a number of products are presented in Rothwell and Zegveld (1979) and Northcott (1980).
5. By the neologism 'electronicisation' we mean the use of electric currents in computers and telecommunication systems to transmit and manipulate information rather than as a medium for energy transmission. The 'mechanisation of information transformation and transfer' would in fact be more consistent with the terminology we develop later in this chapter, so long as 'mechanisation' is not restricted merely to historically 'mechanical' processes but is used to embrace electronic-based 'machines' as well. We use the term 'electronicisation' to show up the parallels between the current application of information technologies to information-handling production processes and traditional applications of 'machines' to manufacturing processes.

6. The figures are from the British Robot Association's 1983 'census' of electronically programmable robots, reported in *New Scientist*, 23 February 1984. The full figures are Japan, 16 500; USA, 8000; West Germany, 4800; France, 2600; Sweden, 1900; Italy, 1800; Britain, 1750; Belgium, 500; Spain, 400; Australia, 300; Finland, 120. The 'census' did not cover robot usage in the state socialist countries.

7. For reviews of the industrial application of robots, see Engelberger (1980), Simons (1980), Marsh (1982) and Industrial Robot (1983).

8. A review of likely technological developments in process industries is in National Economic Development Office: Process Plant Economic Development Committee (NEDO/PPEDC) (1979, 1981). Studies of the rate of diffusion of micro-electronics into all sectors of manufacturing industry in Britain are summarised by Bessant, Braun and Moseley in Forester (1980, pp. 198–218), by Green, Coombs and Holroyd (1980), Braun and Senker (1982), Coppin (1983) and Northcott and Rogers (1984).

9. For developments in steel-making technologies see NEDO/PPEDC (1981); in printing see Martin (1981), Cockburn (1983) and Marshall (1983); in food manufacturing, see Shutt and Leach (1984) and Dickson in Forester (1980, pp. 174–83).

10. New coal-mining technologies are described in Stokes (1982) and National Coal Board (no date); for a critical review of the effects of computerised mining on miners' jobs and working conditions see Burns *et al.* (1983).

11. Reviews of the significance of technological developments in banking are to be found in Bequai (1981) and Marti and Zeilinger (1982).

12. For reviews of such developments see Coombs and Green (1981) and Gershuny and Miles (1983).

3 MECHANISATION AND THE LABOUR PROCESS FROM 1850

1. For accounts of industrialisation from the mid-nineteenth century see Mumford (1934), Hobsbawm (1969, 1975), Landes (1969), Pollard (1981) and Musson (1978).

2. An alternative way of stating this is that a consistent increase in productivity of labour is central to the ability of the economy to follow an equilibrium growth path.

3. This section is based on material from sources listed in Note 1 (Chapter 3) drawing principally on Landes (1969).

4. There were exceptions, such as in some branches of the armaments industry which – for example in the USA – had been using interchangeable parts since the 1820s. Such developments were called by their British admirers the 'American system of manufactures'; see Smith (1977).

5. It is indeed curious that, even though 'Fordism' has been used by many writers to describe aspects of production organisation from the 1920s,

there has been no attempt to define it more rigorously. The principal sources are, of course, the writings of Henry Ford himself (1922, 1929). A fascinating contemporary analysis of Fordism's (or 'Americanism's') political significance is by Gramsci (in Hoare and Nowell Smith, 1971, pp. 279–388). Other discussions of Fordism are Sohn-Rethel (1978), Palloix (1976), Sabel (1982) and Dunford and Perrons (1983).

6. From 1929 to 1939 the proportion of houses connected to electricity supply rose from 20 per cent to 70 per cent in Britain and from 45 per cent to 90 per cent in Germany (Hannah, 1979). From 1930 to 1948 the proportion of consumer expenditure on domestic electrical appliances went up nearly eightfold in Britain; by 1948, 86 per cent of households possessed an electric iron, 64 per cent an electric fire and 40 per cent a vacuum cleaner. The market for electrically-powered consumer durables between the wars in Britain and the USA is discussed in Corley (1966) and Cowan (1976). The rise of new industries in the inter-war period is described in Aldcroft (1970), Pollard (1969), Cook and Stevenson (1979) and Dunford and Perrons (1983).

7. See Beynon (1973), Rothschild (1973), Linhart (1981) and Kamata (1983) for descriptions of conflict in British, American, French and Japanese car plants. Though no less real, the conflicts can be less dramatic in other industries; see Pollert (1981), Herzog (1980) and Nichols and Beynon (1977).

8. Aglietta seems to be either ill-informed or else to have chosen unfortunate terminology in his description of this hardware; he refers to DNC as a transfer line (1979: p. 125). His grasp of the consequences of mechanised control is accurate, however.

9. Sabel (1982) discusses changes in some high volume industries stimulated by changing markets and the arrival of more flexible technologies.

4 MECHANISATION AND LONG WAVES

1. In our discussions in this chapter we assume some knowledge of the basic evidence for long waves (see van Duijn (1983) for a review); recent discussions and elaborations of long wave theories are in Freeman (1983a). Long-wave theories are far from being accepted by all economists (see Maddison, 1982 and Beenstock, 1983). Scepticism arises from the statistical problems of testing the theory; and, one might suggest, from the long-standing project of neo-classical economists to elaborate a theory of *equilibrium* growth. Such a theory might possibly tolerate some fluctuations of an exogenous character in the time-path of the equilibrium, but would not tolerate endogenous cyclical fluctuations of a long duration. There are other problems in establishing the validity of the long wave which are cogently discussed by Rosenberg and Frishstack (1983).

2. In trying to differentiate 'radical' from 'incremental' innovations, Mandel is perhaps making too sharp a distinction. Recent empirical work on the history of particular technological innovations suggests a

smoother, though not continuous, gradation of innovation types; see Freeman (1982) and Abernathy and Utterback (1975).

3. Rowthorn (1976) has criticised Mandel's model; he has pointed out the imprecise nature of Mandel's notion of 'idle capital piling up' from cycle to cycle in anticipation of radical investment projects in the further future. The practical question of the actual location and forms of this capital is not clearly dealt with by Mandel. Is it money-capital (in banks) or is it temporarily under-valued assets which are later realised in the context of enhanced profit expectations?

4. The resemblance of Mandel's approach to the 'random shocks' approach of Maddison (1982) is more apparent than real. For Mandel, exogenous factors are important mainly at the lower turning point but for Maddison, who sees no evidence of waves at all, long depressions *and* long booms are explained by much more contingent factors.

5. There are great similarities with the model of the relationship between innovation and industrial development proposed by Abernathy and Utterback (1975) who note that the development of an industrial unit on the basis of one radical innovation progressively creates obstacles to any other radical innovation in that unit.

6. Freeman points out that not all technologies might follow this sequence and, despite the supposed general world synchronisation of the long wave, particular technologies might be following different time-paths in different countries. He cites as an example the difference in time between the booms in the US and European car industries.

7. Most of the pitfalls concern the manner in which the product titles are chosen, and become embedded in the Standard Industrial Classification scheme which is used for the presentation of Census of Production data. Within any particular branch of the classification scheme, a sub-category such as, say, machine tools will of course encompass a large variety of actual machines. If this category were not broken down further, it would be of little use for the present purpose. Fortunately, it is broken down into many sub-categories. The choice of these further sub-divisions is partly dictated by some obvious production-sequence discontinuity, such as that between drilling machines and milling machines, and partly by the possibility of some identifiable group of machines achieving a particularly significant fraction of production. In general both these criteria have to be present before the census organisers revise their categories to incorporate a new sub-division. Revisions are anyway conducted only infrequently, and the achievement of these criteria is dependent on the product being by common consent an 'established' product.

Therefore, the Census categories themselves are a double-edged sword as indicators of technical change. On the one hand, the appearance and gradual rise to significant fractional share of a new category in the Census is a very reliable indicator of the middle and later portions of the diffusion curve of that product. On the other hand, there can be a significant and randomly variable delay between the real appearance of the new product and its appearance as a census category. Therefore, the new categories might be very poor guides to the very early portions of their diffusion curves. This feature of the way in which the categories of the Census evolve is therefore a major constraint to be borne in mind when using their data for a purpose such as the present one.

A further problem lies in the actual nomenclature of the Census categories and its interpretation. The term chosen by the Census to describe two closely related types of machine may be ambivalent with respect to the technical property in which we happen to be interested. So, while 'pressure vessels' made in the 'fabricated platework' industry will be quite likely to find their way into continuous flow in production industries, 'tanks' made in the same industry will presumably be used in the variety of applications beyond our particular concern of continuous flow production; and there is no information in the Census to help to determine the distribution of tanks between various applications. Here we are forced to use all of one category in our calculations and none of the other, when neither assumption is likely to be completely accurate.

The last problem with the data is that of measurement. Our objective is to identify that fraction of machinery which falls into our several categories of mechanisation and to express its quantity as some sort of proportion to the total amount of machinery, of all sorts, produced in each Census year. In this way we hope to generate a time-profile of the process of mechanisation as revealed by the structure of the output of the capital goods industries. But how are we to express the proportion? If we work in terms of *numbers* of machines we encounter enormous problems of incommensurability. It is hardly appropriate to count both the complete transfer line and a single lathe as one unit of account, yet there is no simple conversion factor. It seems obvious then to use value as a yardstick. The problem here is that the relative price of the new commodity generally changes in a manner different from a general trend of price movements in any period. Therefore, £1m of NC machine tools will not bear the same relation to the total output of the machine tool industry in 1963 as in 1965 for example. The significance of this complication varies depending upon what one is looking for from the data. In one sense it represents useful information, since the high relative prices of new goods at the beginning of their life-cycles contain in part the temporary surplus profits which are the very stimulus to production which is associated with cyclical theories of new product generation. Thus the potential 'inflation' of our measure of the volume of mechanised equipment is not simply 'false', it has a complex significance. For further discussion of these problems, see Coombs (1982).

8. It is nevertheless interesting to note that the mathematical and analytical foundations of Control Theory, so fundamental to the later development of practical electronic control devices were given a large push in the early 1930s by Nyquist and Hazen. Modern control theorists regard their work as vital for the transition from what they call mere 'sequence controlled mechanisation' to 'genuine automation' (Healey, 1975: chap. 1). This distinction corresponds to the terminology of secondary mechanisation as compared with control or tertiary mechanisation used in this book.

5 NEW TECHNOLOGY AND NEO-FORDISM

1. See Burawoy (1979), Cressey and MacInnes (1980), Edwards (1979), Elger (1979), Jones (1982a), Littler (1982), Littler and Salaman (1984) and Thompson (1983).
2. For clerical work see Braverman (1974) Part IV, Barker and Downing (1980), de Kadt (1979), Glenn and Feldberg (1979), Crompton and Reid (1982); for computer programming, see Kraft (1977) and Greenbaum (1976).
3. See Buchanan (1979) for a review of this field.
4. See Engelstad (1979) for paper-making; Walton (1972, 1977) for food-processing; Hill (1971) for oil-refining; Roeber (1975) for chemical production.
5. See Braverman (1974: p. 38), Hales (1974), Zimbalist (1979), Nichols and Beynon (1977) and Rinehart (1978).
6. See Kelly (1978) and reviews in Buchanan (1979), Herbst (1974), van der Zwaan (1975), Rose (1975) and Brown (1967).
7. See Hulin and Blood (1968), Pierce and Dunham (1976), Cummings and Molloy (1976), Cummings *et al.* (1977) and Srivastva *et al.* (1975).
8. As writers on segmentation in labour markets have noted, often the core and peripheral roles draw on different sectors of the workforce; see Gordon *et al.* (1982) for a discussion of segmentation in American industry.
9. See Thompson (1967), Williamson (1975), Galbraith (1977) and Perrow (1979).
10. Of interest here is the fact that the realisation of such work intensification and responsibility for quality was dependent on such factors as the existence of individualised payment systems and a reduction of the role of ancillary workers. Kelly argues that this can be held to account for productivity increases with more plausibility than can the effects of 'satisfaction' postulated by job redesign theorists.
11. For example, see Hackman and Oldham (1980) and Lupton (1975).
12. See examples from the 'micro-electronics revolution' debate in Green (1984).
13. Thus it is not clear that level 7 (remote control of a power tool) is more 'mechanised' than level 6 (fixed sequence control over power tool). Similarly, is the mechanisation of workpiece insertion on a fixed sequence machine more mechanised than a hand-fed machine with some kind of workpiece measurement capability? Bell's 3-D model of mechanisation with independent scales avoids this problem in Bright's scalar classification.
14. For example see Nichols and Beynon (1977) for a description of the heavy manual work which can still exist within a supposedly automated chemical plant.
15. Examples are: the Digitron computer-controlled system and information system running parallel to the flexible pallet (Robogate) system in Fiat (Amin, 1982); the information system designed for use in the Volvo Kalmar plant (Aguren *et al.*, 1976).

16. An exception is Sabel (1982) who discusses the work organisation aspects of neo-Fordism in the context of the 'end of Fordism'. He draws his examples from a number of industries, almost entirely in manufacturing, seeking to show that even Fordist mass production firms are required to respond to the demands of flexibility. Though neo-Fordism can be discussed in relation to existing Fordist-organised industries, we prefer to emphasise its special significance for the sectors we discuss in Chapters 6 and 7.

17. See Nygaard and Fjalestad (1981), Mumford (1982), Rosenbrock (1979) and Council for Science and Society (1981).

6   NEO-FORDISM IN SMALL-BATCH ENGINEERING

1. In the case of skilled labour shortages much of the blame must lie with inadequate levels of apprentice training; see Engineering Industry Training Board (1975), National Economic Development Office (1980) and Noble (1979: p. 42).

2. Reviews of GT appear in Edwards (1971), Gallagher and Knight (1973), National Economic Development Office (1975) and Burbidge (1975). Examples of it are described in Ranson (1972), Marklew (1970, 1972a, 1972b) and Craven (1974). Criticisms of these expositions of GT can be found in Bornat (1978) and Green (1978).

3. This is especially so in the case of British metal-cutting craft workers, as described by Penn (1982) and More (1980).

4. If skilled workers carry out the handling and loading tasks, the skilled content of their work is likely to be increased by the removal of these less-skilled transfer functions. It is possible that this may result in the displacement of some skilled workers assuming no increase in output, yet given the volume of work necessary to amortise any such investment in transfer technology such effects are likely to be of minor significance at present. Much more common is the displacement of less skilled labour performing such tasks.

5. In April 1984, the Japanese Agency of Industrial Science and Technology opened a 'dream factory of the twenty-first century' in Tsukuba, near Tokyo, in which 'the only human roles are inspection and sweeping the floor'. (*The Guardian*, 17 April 1984, p. 25). The *Financial Times*, 3 November 1984, p. 9, however, reports that an earlier version of this factory had not proved itself economically. Although the number of employees had been reduced from 215 to 12 and the processing time from 35 to 1.5 days the financial return over the first two years was less than $7m. on an investment of $18m. What is more, a survey conducted by Ingersoll Engineers showed that only two or three of the fifty Flexible Manufacturing Systems in Japan, Europe and the USA seemed financially viable. For a review of Japanese developments in 'mechatronics' ('the combination of micro-electronics and mechanical engineering') by Japanese and European engineers and industrialists see McLean (1983). Havatny (1983) surveys existing FMSs in Japan, the USA and Europe (both Western and Eastern).

7 INFORMATION TECHNOLOGY, THE SERVICE
   SECTOR AND THE RESTRUCTURING OF
   CONSUMPTION

1. See, for example, Bell (1974), Porat (1977), Stonier (1983) and Barry
   Jones (1982b). For a critique of the various classificatory schemes see
   Kumar (1978).
2. See Bell (1979).
3. It should be said that voice-driven electronic equipment is in the very
   earliest stages of its diffusion. A 'voice-driven financial spreadsheet'
   software package is now on sale in Europe (at £1000) with which 'a user
   can give commands, enter data, perform "what if?" analyses, insert or
   delete rows of information . . . just by talking to the system through a
   microphone'. (*Computing*, 12 April 1984, p. 3).
4. And even in the late 1970s office workers had an average of only one
   tenth of the capital equipment of shop-floor workers (TUC, 1979: p. 24).
5. See Hoos (1961), Stymne (1966), Mumford and Banks (1967), Whisler
   (1970) and Argyris (1973).
6. Such a view has been argued thus:

   > While there has been a tendency historically for the secretarial
   > worker's job to develop as a completely manual function, many
   > secretarial workers find they have auxiliary tasks to perform, such as
   > filing, the responsibility for an individual's mail and making coffee.
   > The object of word-processing systems is to fragment the secretarial
   > job further. High levels of efficiency and cost-effectiveness are best
   > achieved by stripping away all possible distractions from the actual job
   > of typing . . . Proponents of word-processing argue that the new
   > technology will lighten the burden of office work and free secretarial
   > workers for more interesting work. Perhaps the top secretaries will be
   > given a more administrative function, but the largely working-class
   > women who are on the lower levels of the hierarchy will find their
   > work becoming more boring, routine and intensified. (Conference of
   > Socialist Economists Micro-electronics Group, 1980: pp. 48–9)

7. See McCabe and Popham (1977), Bird (1980) and West (1981).
8. For more on retailing technologies see Marti and Zeilinger (1982) and
   Green *et al.* (1983).
9. Aglietta describes Fordism rather indigestibly as 'the principle of an
   articulation between process of production and mode of consumption
   which constitutes the mass production that is the specific content of the
   universalisation of wage labour' (1979: p. 117).
10. Aglietta lists the 'limits to Fordism' as applied to the mass production
    sectors thus:

    • the synchronisation of production sequences to keep labour and
      capital working all the time; this includes such problems as the time
      taken to start up and balance the work between workers on the
      production line.

- the limits of speed-up imposed by the 'nervous exhaustion' of workers.
- the difficulties of tying individual effort levels to collective output within a collective means of production. As a result there are attempts to work collective bonus schemes, with measured day work replacing piecework, and to introduce 'worker participation' against the collective resistance of certain groups of workers. This was less a problem in post-war Europe as workers' collective power had been weakened by fascism and war but as trade unions organised through the 1950s and 1960s, the flexibility of task allocation became more difficult. This was reflected in the late 1960s and early 1970s in the literature concerned with higher absenteeism, labour turnover, strikes and sabotage in Fordist industries.

These problems facing firms in the mass production sector have not yet been solved, although there are a number of means being sought to solve them. Technological developments, in computer-aided management and in robotics for example, provide means of coping with synchronisation problems and with the resistance of organised workers by reducing their crucial role in the production process. Either by different forms of work organisation (autonomous work groups, quality circles) by political action (legislation restricting workers' trade union power) or by economic pressure (through higher unemployment or international movement of new investment) the forms of worker veto over the reorganisation of production on more profitable lines can be circumvented.

11. Thus, in 1954, 60 per cent of the average British household expenditure on entertainment was spent on cinema, theatre, etc. admissions and 40 per cent on the purchase and hire of televisions, radios, etc. By 1974 the balance had shifted, to 21 per cent and 79 per cent respectively. A similar change can be observed in the balance between domestic 'help', laundry, etc. and domestic appliance purchase (1954: 67 per cent to 33 per cent; 1974: 32 per cent to 68 per cent). In transport the shift was even greater from services to goods (1954: 50 per cent to 50 per cent; 1974: 18 per cent to 82 per cent). See Gershuny (1983: p. 18).

12. This is an effect only partly realised in practice because of the elastic nature of domestic labour requirements; see Huws (1982: p. 92).

13. In Britain, such convictions have been a continual theme of the post-1979 Conservative government, although the rhetoric has often concealed the practical difficulties of radical reductions in state spending; see Hall and Jacques (1983) and Riddell (1983).

14. For accounts of some developments, actual and possible, see Hastings and Levie (1983) on recent 'privatisation' attempts in Britan; Gershuny and Miles (1983: pp. 85–91) on the Open University as an alternative, cheaper system of higher education and Jones, C. (1983) on computerising social work.

15. The British government departments of Health and Social Security and of Trade and Industry are involved in a number of projects developing appropriate systems for computerising health records in hospitals and

general practice. See *Computing*, 16 February 1984, pp. 20–1 and 22 March 1984, pp. 24–5; For a general discussion of computers and medical records see Kember (1982) and Bronzino (1982).

16. In the USA, where private medicine accounts for the majority of medical care, one third of operations are carried out via day surgery. Yet in Britain only 20 per cent of private medicine takes place on a day-care basis (*Financial Times*, survey, 24 January 1984).

17. For details, see Rogers *et al.* (1979), Bronzino (1982), Kember (1982) and Miller *et al.* (1982). The last two have extensive bibliographies. See also Child *et al.* (1983).

18. For details of these devices see *Venture*, June 1982, p. 42.

19. See *Computing*, 9 February 1984, pp. 24–5.

20. Reported in North West Industrial Development Association (1983).

21. See *Financial Times*, survey, 14 January 1982.

22. The British health insurance organisation, BUPA, runs an off-the-street health check clinic in London, at £150 a 'consultation'.

23. *Practical Computing*, April 1984, pp. 74–8 reviews the medical, keep fit and first-aid software on sale in Britain for home computer owners.

8 CONCLUSION: FROM FORDISM TO NEO-FORDISM?

1. Perez (1983) has made a similar point, although with less historical detail, seeing boom periods of expansion as indicative of a 'good match' between new technological systems and the socio-institutional arrangements. Depressions are thus periods of mis-match between these two.

2. Such arguments have been put forward recently by Hales (1980) and by a number of liberal engineers; see Rosenbrock (1979), Council for Science and Society (1981) and Thring (1980). The engineering research that their ideas suggest would require more substantial state support than is at present forthcoming, and could be part of some 'alternative' research programme enacted by a radical government of the left. (See Group for Alternative Science and Technology Strategies, 1982.)

3. This of course has been the subject of much study by geographers seeking to describe recent locational changes in industrial production; see Massey and Meegan (1982) and Fothergill and Gudgin (1982). These locational factors – the shift of industry to the US 'sunbelt' or the UK M4 corridor – have important political implications; for the UK, see Massey (1983) and for the USA, see Davis (1984).

4. Murray (1983), using examples from Italian industry, discusses the influence of new electronic-based technologies on the location of production units. He emphasises how managements can use such technologies to gain greater control over production and weaken the constraints placed on their prerogatives to organised workers.

5. These points are discussed further in Blackburn, Green and Liff (1982); for a preliminary assessment of the innovation and employment strategies of the Greater London Council see Elliott (1984).

Bibliography

Abernathy, W. J. (1978) *The Productivity Dilemma* (Baltimore: Johns Hopkins University Press).

Abernathy, W. J. and Utterback, J. (1975) 'A dynamic model of process and product innovation', *Omega*, vol. 3, no. 6, pp. 639–56.

Aglietta, M. (1979) *A Theory of Capitalist Regulation: the US Experience* (London: New Left Books).

Aguren, S., Hannsson, R. and Karlsson, K. G. (1976) *The Volvo Kalmar Plant* (Stockholm: Rationalization Council).

Aldcroft, D. H. (1970) *The Inter-War Economy: Britain 1919–1939* (London: Batsford).

Alvey Committee (1982) *A Programme for Advanced Information Technology* (London: HMSO).

Amin, A. (1982) 'Restructuring in Fiat and Decentralisation of Production into Southern Italy', mimeo.

Argyris, C. (1973) *On Organisations of the Future* (London: Sage).

Arnold, E. and Senker, P. (1982) *Designing the Future – the Implications of CAD Interactive Graphics for Employment and Skills in the British Engineering Industry*, Occasional Paper 9 (Watford: Engineering Industry Training Board).

Atkinson, J. (1980) 'The Place of the Industrial Robot in the Evolution of Automation', unpublished M.Sc. thesis, University of Manchester Institute of Science and Technology.

Atkinson, J. (1984) 'Flexible Firm Takes Shape', *The Guardian*, 18 April, p. 19.

Automated Small Batch Production Committee (1978) *Automated Small Batch* (Glasgow: National Engineering Laboratory).

Ayres, A. U. and Miller, S. (1981) 'Robotics, CAM and Industrial Productivity', *National Productivity Review*, vol. I, no. 1, pp. 42–60.

Barker, J. and Downing, H. (1980) 'Word Processing and the Transformation of the Patriarchal Relations of Control in the Office', *Capital and Class*, no. 10, pp. 64–9.

Barron, I. and Curnow, R. (1979) *The Future with Micro-electronics*, (London: Frances Pinter).

Beenstock, M. (1983) *The World Economy in Transition* (London: Allen & Unwin).

Bell, D. (1974) *The Coming of Post-Industrial Society: A Venture in Social Forecasting* (London: Heinemann).

Bell, D. (1979) 'The Social Framework of the Information Society', in Dertouzous and Moses (1979) and in Forester (1980).

Bell, R. M. (1972) *Changing Technology and Manpower Requirements in the Engineering Industry* (Brighton: Sussex University Press).

Benet, M. K. (1972) *Secretary* (London: Sidgwick & Jackson).

Bequai, A. (1981) *The Cashless Society: EFTS at the Crossroads* (New York: Wiley–Interscience).

Best, M. H. and Connolly, W. E. (1982) *The Politicized Economy*, 2nd ed (Lexington: D. C. Heath).

Beynon, H. (1973) *Working for Ford* (Harmondsworth: Penguin).

Bird, E. (1980) *Information Technology in the Office: the Impact on Women's Jobs* (Manchester: Equal Opportunities Commission).

Blackburn, P., Green, K. and Liff, S. (1982) 'Science and Technology in Restructuring', *Capital and Class*, no. 18, pp. 15–37.

Blau, P. M., *et al.* (1976) 'Technology and Organisation in Manufacturing', *Administrative Science Quarterly*, vol. 21, no. 1, pp. 20–40.

Blauner, R. (1964) *Alienation and Freedom* (Chicago: University of Chicago Press).

Blumberg, M. and Gerwin, D. (1981) 'Coping with Advanced Manufacturing Technology', Discussion Paper 81–12 (Berlin: International Institute of Management).

Boon, J. (1981) 'The Development of Operator Programmable NC Lathes', mimeo.

Bornat, A. (1978) 'Group Technology and Group Working', unplublished M.Sc. dissertation, University of Manchester.

Braun, E. and Senker, P. (1982) *New Technology and Employment* (London: Manpower Services Commission).

Braverman, H. (1974) *Labor and Monopoly Capital: The Degradation of Work in the Twentieth Century* (New York: Monthly Review Press).

Bright, J. R. (1958) *Automation and Management* (Boston: Harvard Business School).

Brighton Labour Process Group (1977) 'The Capitalist Labour Process', *Capital and Class*, no. 1, pp. 3–26.

Bronzino, J. D. (1982) *Computer Applications for Patient Care* (Menlo Park, California: Addison-Wesley).

Brown, R. K. (1967) 'Research and Consultancy in Industrial Enterprises', *Sociology*, vol. 1, pp. 33–60.

Buchanan, D. (1979) *The Development of Job Design Theories and Techniques* (Farnborough: Saxon House).

Buckingham, W. S. (1955) in United States Congress (1955).

Burawoy, M. (1979) *Manufacturing Consent: Changes in the Labour Process under Monopoly Capitalism* (Chicago: University of Chicago Press).

Burbidge, J. L. (1975) *The Introduction of Group Technology* (London: Heinemann).

Burbidge, J. L. (1978) 'Whatever Happened to GT?', *Management Today*, September, pp. 87–9.

Burns, A., Feickert, D., Newby, M. and Winterton, J. (1983) 'The Miners and New Technology', *Industrial Relations Journal*, vol. 4, Winter, pp. 7–20.

Cavendish, R. (1982) *Women on the Line* (London: Routledge & Kegan Paul).

Chandler, A. D. (ed.) (1964) *Giant Enterprise* (New York: Harcourt, Brace & World).

Child, J. *et al.* (1983) 'Micro-electronics and the Quality of Employment in Services', in Marstrand (1984).

Clark, J., Freeman, C. and Soete, L. (1981) 'Long Wave, Inventions and Innovations' in Freeman (1983a).

Cockburn, C. (1983), *Brothers: Male Dominance and Technological Change* (London: Pluto Press).

Cohen, J. (1975) 'The Automation Debate in Perspective', unpublished M.Sc. dissertation, Manchester University.

Conference of Socialist Economists (1976) *The Labour Process and Class Strategies* (London: Stage One).

Conference of Socialist Economists Micro-electronics Group (1980) *Micro-electronics: Capitalist Technology and the Working Class* (London: CSE).

Cook, C. and Stevenson, J. (1979) *The Slump: Society and Politics during the Depression* (London: Quartet).

Cook, N. H. (1975) 'Computer-managed Parts Manufacture', *Scientific American*, February, pp. 23–8.

Cooley, M. (1980) *Architect and Bee: the Human/Technology Relationship* (Slough: Hand and Brain).

Coombs, R. (1981) in Freeman (1983a).

Coombs, R. (1982) 'Automation and Long Wave Theories', unpublished Ph.D. thesis, Manchester University.

Coombs, R. (1984) 'Automation, Management Strategies and Labour Process Change' in Knights (1984).

Coombs, R. and Green, K. (1981) 'Micro-electronics and the Future of Service Employment', *Service Industries Review*, no. 2, July, pp. 4–21.

Coppin, P. (1983) *New Technology: Manufacturing*, Research Report 56, (London: Hammersmith and Fulham Development Planning Department).

Coriat, B. (1980) 'The Restructuring of the Assembly Line: a New Economy of Time and Control', *Capital and Class*, no. 11, pp. 33–43.

Corley, T. A. B. (1966) *Domestic Electrical Appliances* (London: Jonathan Cape).

Council for Science and Society (1981) *New Technology: Society, Employment and Skill* (London: CSS).

Cowan, R. S. (1976) 'The Industrial Revolution in the Home: Household Technology and Social Change in the Twentieth Century', *Technology and Culture*, vol. 17, pp. 1–23.

Craven, F. W. (1974) 'GT Applications', *Machine Tool Review*, March, pp. 48–55.

Cressey, P. and MacInnes, J. (1980) 'Voting for Ford: Industrial Democracy and the Control of Labour', *Capital and Class*, no. 11 pp. 5–33.

Crompton, R. and Reid, S. (1982) in Wood, S. (1982).

Crum, R. E. and Gudgen, G. (1978) *Non-Production Activities in UK Manufacturing Industry* (Brussels: European Economic Community).

Cummings, T. G. and Molloy, E. S. (1976) *Improving Productivity and the Quality of Working Life* (New York: Praeger Books).

Cummings, T. G. *et al.* (1977) 'A Methodological Critique of 58 Selected Work Experiments', *Human Relations*, vol. 30, no. 8, pp. 675–708.

Dale, J. R. (1962) *The Clerk in Industry* (Liverpool: Liverpool University Press).

Davis, L. and Taylor, J. C. (eds) (1979) *Design of Jobs,* 2nd ed (Santa Monica: Goodyear).

Davis, M. (1984) 'The Political Economy of Late Imperial America', *New Left Review,* no. 143, January–February, pp. 6–37.

Dawson, J. A. (1981) 'Innovation Adoption in Food Retailing: the Example of Self-service Methods', *Service Industries Review,* vol. 1, no. 2, pp. 22–35.

Day, R. (1981) *The 'Crisis' and the 'Crash': Soviet Studies of the West, 1917–1939* (London: New Left Books).

Delbeke, J. (1981) 'Recent Long-wave Theories: a Critical Survey' in Freeman (1983a).

Dertouzos, M. L. and Moses J. (eds) (1979) *The Computer Age: A Twenty Year View* (Cambridge: Massachusetts Institute of Technology Press).

Doring, M. R. and Salling, R. C. (1971) 'A Case for Wage Incentives in the NC Age', *Manufacturing Engineering Management,* vol. 66, pp. 31–3.

Dosi, G. (1983) 'Technology and Conditions of Macro-economic Development', paper presented for the workshop on 'Innovation, Design and Long Cycles in Economic Development', Royal College of Art, London, April.

Doyal, L. (1979) *The Political Economy of Health* (London: Pluto Press).

Duijn, J. J. van (1983) *The Long Wave in Economic Life* (London: Allen & Unwin).

Dunford, M. and Perrons, D. (1983) *The Arena of Capital* (London: Macmillan).

Edwards, G. A. B. (1971) *Readings in Group Technology* (Brighton: Machinery Publishing Company).

Edwards, G. A. B. (1974a) 'Group Technology: a Technical Answer to a Social Problem', *Personnel Management,* March, pp. 35–9.

Edwards, G. A. B. (1974b) 'Group Technology – Technique, Concept or New Manufacturing System?', *The Chartered Mechanical Engineer,* vol. 21, no. 2, pp. 67–70 and no. 3, pp. 61–6.

Edwards, G. A. B. (1980) 'What's Wrong with British Factories?', *Management Today,* February, pp. 58–61.

Edwards, R. (1979) *Contested Terrain: the Transformation of the Workplace in the Twentieth Century* (London: Heinemann).

Elger, T. (1979) 'Valorisation and "Deskilling": a Critique of Braverman', *Capital and Class,* no. 7, pp. 58–99.

Elkington, J. (1984) *Sun Traps: the Renewable Energy Forecast* (Harmondsworth: Penguin).

Elliott, D. (1984) *The GLC's Innovation and Employment Initiatives,* Occasional Paper 7 (Milton Keynes: Open University Technology Policy Group).

Engelberger, J. F. (1980) *Robotics in Practice* (London: Kogan Page).

Engelstad, P. H. (1979) 'Socio-technical Approach to Problems of Process Control' in Davis, L. and Taylor, J. C. (1979).

Engineering Industry Training Board (1975) *The Craftsman in Engineering* (Watford: EITB).

European Community Commission (1983) *Europe 1995: Technological Change and Social Impact* (Paris: Futuribles).

Evans, C. (1979) *The Mighty Micro* (London: Victor Gollancz).

Feigenbaum, E. A. and McCorduck, P. (1984) *The Fifth Generation: Artificial Intelligence and Japan's Computer Challenge .o the World* (London: Pan).

Foley, G. (1981) *The Energy Question* (Harmondsworth: Penguin).

Ford, G., Georghiou, L. and Cameron, H. (1983) 'Using the Seas During the Twenty-first Century', *Impact of Science on Society*, no. 334, pp. 491–501.

Ford, H. (1922) *My Life and Work* (New York: Doubleday Page).

Ford, H. (1929) 'Mass Production', *Encyclopaedia Brittanica* (1929).

Forester, T. (ed.) (1980) *The Micro-electronics Revolution: the Complete Guide to the New Technology and its impact on Society* (Oxford: Basil Blackwell).

Fothergill, S. and Gudgin, G. (1982) *Unequal Growth: Urban and Regional Employment Change in the UK* (London: Heinemann).

Freeman, C. (1977) 'The Kondratiev Long Waves, Technical Change and Unemployment' in Organisation for Economic Co-operation and Development (1979).

Freeman, C. (1982) *The Economics of Industrial Innovation*, 2nd ed (London: Frances Pinter).

Freeman, C. (ed.) (1983a) *Long Waves in the World Economy* (Sevenoaks: Butterworths).

Freeman, C. (1983b) 'Keynes or Kondratiev? How Can we Get Back to Full Employment' in Marstrand (1984).

Freeman, C., Clark, J. and Soete, L. (1982) *Unemployment and Technical Innovation* (London: Frances Pinter).

Friedman, A. (1977) *Industry and Labour: Class Struggle at Work and Monopoly Capitalism* (London: Macmillan).

Friedrichs, G. and Schaff, A. (eds) (1982) *Micro-electronics and Society: For Better or For Worse* (Oxford: Pergamon).

Galbraith, J. R. (1977) *Organisational Design* (Reading, Massachusetts: Addison-Wesley).

Gallagher, C. C. and Knight, W. A. (1973) *Group Technology* (Sevenoaks: Butterworths).

Gershuny, J. (1983) *Social Innovation and the Division of Labour* (Oxford: Oxford University Press).

Gershuny, J. and Miles, I. (1983) *The New Service Economy: the Transformation of Employment in Industrial Societies* (London: Frances Pinter).

Gerwin, D. and Leung, T. K. (1980) 'The Organisational Impacts of FMS: Some Initial Findings', discussion paper, Institute for Social Research in Industry, Trondheim, Norway.

Gibbons, M. and Gummett, P. (eds) (1984) *Science, Technology and Society Today* (Manchester: Manchester University Press).

Glenn, E. N. and Feldberg, R. L. (1979) 'Proletarianising Clerical Work: Technology and Organisational Control in the Office' in Zimbalist (1979).

Gordon, D. M., Edwards, R. and Reich, M. (1982) *Segmented Work, Divided Workers: the Historical Transformation of Labour in the United States,* (Cambridge: Cambridge University Press).

Gough, I. (1979) *The Political Economy of the Welfare State* (London: Macmillan).

Green, K. (1978) 'Group Technology in Small Batch Engineering', paper given at Windsor Deskilling Conference, 18–19 December.

Green, K. (1984) in Gibbons and Gummett (1984).

Green, K., Coombs, R. and Holroyd, K. (1980) *The Effects of Micro-electronic Technologies or Employment Prospects: a Case Study of Tameside* (Farnborough: Gower).

Green, K., Liff, S., Taylor, A. and Wield, D. (1983) 'New Technology in Retailing and Distribution', mimeo.

Greenbaum, J. (1976) 'Division of Labour in the Computer Field', *Monthly Review,* vol. 28, no. 3, pp. 40–55.

Group for Alternative Science and Technology Strategies (1982) *Redirecting Science and Technology,* mimeo.

Hackman, J. R. and Oldham, G. R. (1980) *Work Redesign* (Reading, Massachusetts: Addison-Wesley).

Hales, M. (1974) 'Management Science and the Second Industrial Revolution', *Radical Science Journal,* no. 1, pp. 5–28.

Hales, M. (1980) *Living Thinkwork: Where do Labour Processes Come from?* (London: CSE Books).

Hall, S. and Jacques, M. (eds) (1983) *The Politics of Thatcherism,* (London: Lawrence & Wishart).

Hannah, L. (1979) *Electricity before Nationalisation* (London: Macmillan).

Hastings, S. and Levie, H. (eds) (1983) *Privatisation?* (Nottingham: Spokesman).

Havatny, J. (1983) *World Survey of CAM* (Sevenoaks: Butterworths).

Healey, M. (1975) *Principles of Automatic Control* (London: English Universities Press).

Herbst, P. G. (1974) *Socio-technical Design* (London: Tavistock).

Herzog, M. (1980) *From Hand to Mouth: Women and Piecework,* (Harmondsworth: Penguin).

Hill, C. T. and Utterback, J. M. (eds) (1979) *Technological Innovation for a Dynamic Economy* (New York: Pergamon).

Hill, P. (1971) *Towards a New Philosophy of Management* (Aldershot: Gower Press).

Hines, C. and Searle, G. (1979) *Automatic Unemployment* (London: Earth Resources Research).

Hoare, Q. and Nowell Smith, G. (eds) (1971) *Selections from the Prison Notebooks of Antonio Gramsci* (London: Lawrence & Wishart).

Hobsbawm, E. (1969) *Industry and Empire: An Economic History of Britain since 1750* (Harmondsworth: Penguin).

Hobsbawm, E. (1975) *The Age of Capital* (London: Weidenfeld & Nicolson).

Holland, S. (ed) (1983) *Out of Crisis: A Project for European Recovery* (Nottingham: Spokesman).

Hollingum J. (1980) 'ASP programme Sets a Survival Course for British Manufacturing', *The Engineer*, 31 January, pp. 33–9.

Hoos, I. R. (1961) *Automation in the Office* (Washington, DC: Public Affairs Press).

Hounshall, D. A. (1981a) 'Mass Production and Consumption in the US 1850–1930', paper presented to a British Society for the History of Science Conference on 'New Perspectives in the History of Technology', Manchester University, 27–29 March.

Hounshall, D. A. (1981b) 'Mass production, Technologists and Work Experience in America', paper presented at the Annual Meeting of the Organisation of American Historians, Detroit, 2 April.

Hulin, C. and Blood, M. R. (1968) 'Job enlargement, individual differences and worker response', *Psychological Bulletin*, vol. 69, pp. 41–55.

Hull, D. (1978) *The Shop Steward's guide to work organisation* (Nottingham: Spokesman).

Huws, U. (1982) *Your Job in the Eighties: a Woman's Guide to New Technology* (London: Pluto Press).

Huws, U. (1984) 'New Technology homeworkers', *Employment Gazette*, January, pp. 13–17.

Iliffe, S. (1983) *The NHS: A Picture of Health?* (London: Lawrence & Wishart).

Industrial Robot Magazine (1983) *Decade of Robotics*, Special Tenth Anniversary Issue (Bedford: IFS Publications).

Jenkins, C. and Sherman, B. (1979) *The Collapse of Work* (London: Eyre Methuen).

Johnston, R. and Gummett, P. (eds) (1979) *Directing Technology* (London: Croom Helm).

Jones, Bryn (1982a) 'Destruction or Redistribution of Engineering Skills? The Case of Numerical Control' in Wood (1982).

Jones, Barry (1982b) *Sleepers, Wake! Technology and the Future of Work* (Brighton: Wheatsheaf).

Jones, C. (1983) *State Social Work and the Working Class* (London: Macmillan).

Kadt, M. de (1979) 'Insurance: a Clerical Work Factory' in Zimbalist (1979).

Kamata, S. (1983) *Japan in the Passing Lane* (London: Allen & Unwin).

Kelly, J. E. (1978) 'A Reappraisal of Socio-technical Systems Theory', *Human Relations*, vol. 31, pp. 1069–99.

Kelly, J. E. (1979) 'Job Redesign: a Critical Analysis', unpublished Ph.D. thesis, University of London, London School of Economics.

Kelly, J. E. (1982a) *Scientific Management, Job Redesign and Work Performance* (New York: Academic Press).

Kelly, J. E. (1982b) 'Economic and Structural Analysis of Job Redesign' in Kelly, J. E. and Clegg, C. W. (eds) (1982) *Autonomy and Control at the Workplace* (London: Croom Helm).

Kember, N. F. (1982) *An Introduction to Computer Applications in Medicine* (London: Edward Arnold).

Kleinknecht, A. (1981a) 'Observations on the Schumpeterian Swarming of Innovations' in Freeman (1983a).

Kleinknecht, A. (1981b) 'Prosperity, Crises and Innovation Patterns: Some More Observations on Neo-Schumpeterian Hypotheses', mimeo.

Knights, D. (ed.) (1984) *Job Redesign: Critical Perspectives in the Labour Process* (London: Heinemann).

Kraft, P. (1977) *Programmers and Managers: The Routinisation of Computer Programming in the US* (New York: Springer Verlag).

Kraft, P. (1979) 'Challenging the Mumford Democrats at Derby Works', *Computing Europe*, 2 August, pp. 17–18.

Kumar, K. (1978) *Prophecy and Progress* (Harmondsworth: Penguin).

Landes, D. S. (1969) *The Unbound Prometheus* (Cambridge: Cambridge University Press).

Leavitt, H. J. and Whisler, J. L. (1958) 'Management in the 1980s', *Harvard Business Review*, vol. 36, November–December, pp. 41–8.

Liff, S. (1982) 'Making Technology Visible', *Scarlet Woman*, no. 14, pp. 15–18.

Likert, R. (1961) *New Patterns of Management* (New York: McGraw-Hill).

Linhart, R. (1981) *The Assembly Line* (London: John Calder).

Lipietz, A. (1982) 'Towards Global Fordism', *New Left Review*, no. 132, pp. 38–58.

Littler, C. R. (1982) *The Development of the Labour Process in Capitalist Societies* (London: Heinemann).

Littler, C. R. and Salaman, G. (1984) *Class at Work: The Design, Allocation and Control of Jobs* (London: Batsford).

Lupton, T. (1975) 'Efficiency and the Quality of Work Life: the Technology of Reconciliation', *Organisational Dynamics*, Autumn, pp. 68–80.

McCabe, H. and Popham, E. (1977) *Word Processing: A Systems Approach to the Office* (New York: Harcourt Brace Jovanovitch).

McLean, M. (1983) *Mechatronics: Developments in Japan and Europe* (London: Frances Pinter).

McLean, M. and Rush, H. J. (1978) 'The Impact of Micro-electronics on the UK', occasional paper series No. 7, Science Policy Research Unit, University of Sussex.

Maddison, A. (1982) *Phases of Capitalist Development* (Oxford: Oxford University Press).

Manchester Employment Research Group (1983) *GEC Power Engineering: New Technology* (Manchester: MERG).

Mandel, E. (1975) *Late Capitalism* (London: New Left Books).

Mandel, E. (1978) *The Second Slump* (London: New Left Books).

Mandel, E. (1980) *Long Waves of Capitalist Development* (Cambridge: Cambridge University Press).

Mandel, E. (1981) in Freeman (1983a).

Manpower Services Commission (1982) *Text Processing*, 4 volumes (London: MSC).

Marglin, S. A. (1974) 'What Do Bosses Do? The Origins and Functions of Hierarchy in Capitalist Production', *Review of Radical Political Economics*, vol. 6, no. 2, pp. 60–112.

Marklew, J. J. (1970) 'Advanced Ideas for Production in the 70s at Ferranti, Edinburgh', *Mechanical and Production Engineering*, 4 February, pp. 162–8.

Marklew, J. J. (1972a) 'An Application of the Cell System in a Small Machine Shop', *Mechanical and Production Engineering*, 23 August, pp. 259–63.

Marklew, J. J. (1972b) 'Cell System of Manufacture at Platt International Ltd', *Mechanical and Production Engineering*, 13 December, pp. 830–3.

Marsh, P. (1981a) 'A Fresh Start for British Machine Tools', *New Scientist*, 26 February, pp. 544–6.

Marsh, P. (1981b) *The Silicon Chip Book* (London: Abacus).

Marsh, P. (1982) *The Robot Age* (London: Abacus).

Marshall, A. (1983) *Changing the Word: the Printing Industry in Transition* (London: Comedia).

Marstrand, P. (ed.) (1984) *New Technology and the Future of Work and Skills* (London: Frances Pinter).

Marti, J. and Zeilinger, A. (1982) *Micros and Money: New Technology in Banking and Shopping* (London: Policy Studies Institute).

Martin, R. (1981) *New Technology and Industrial Relations in Fleet Street* (Oxford: Oxford University Press).

Marx, K. (1976) *Capital, Volume One* (Harmondsworth: Penguin). (First published 1867).

Massey, D. (1983) 'The Shape of Things to Come', *Marxism Today*, April, pp. 18–27.

Massey, D. and Meegan, R. (1982) *The Anatomy of Job Loss: the How, Why and Where of Employment Decline* (London: Methuen).

Mensch, G. (1979) *The Technological Stalemate* (Cambridge, Massachusetts: Ballinger).

Miller, R. A. *et al.* (1982) 'INTERNIST–I, an Experimental Computer-based Diagnostic Consultant for General Internal Medicine', *New England Journal of Medicine*, vol. 307, pp. 468–76.

More, C. (1980) *Skill and the English Working Class 1870–1914* (London: Croom Helm).

Mortensen, N. (1979) 'Impact of Technological and Market Changes in Organisational Functioning: the Conceptual Framework in Bourgeois and Marxist Theory and Research', *Acta Sociologica*, vol. 22, no. 2, pp. 135–59.

Mumford, E. (1982) *Values, Technology and Work* (The Hague: Martinus Nijhoff).

Mumford, E. (1983) *Designing Secretaries: the Participative Design of a Word-processing System* (Manchester: Manchester Business School).

Mumford, E. and Banks, O. (1967) *The Computer and the Clerk* (London: Routledge & Kegan Paul).

Mumford, E. and Henshall, D. (1983) *Designing Participatively: a Participative Approach to Computer Systems Design* (Manchester: Manchester Business School).

Mumford, E., Land, F. and Hawgood, J. (1978) 'A Participative Approach to the Design of Computer Systems', *Impact of Science on Society*, vol. 28, no. 3, pp. 235–53.

Mumford, L. (1934) *Technics and Civilisation* (London: Routlege & Kegan Paul).

Murray, F. (1983) 'Decentralisation of Production – the decline of the Mass Collective Worker?', *Capital and Class,* no. 19, pp. 74–99.

Musson, A. E. (1978) *The Growth of British Industry* (London: Batsford).

Nabseth, L. and Ray, G. F. (eds) (1974) *The Diffusion of New Industrial Processes* (Cambridge: Cambridge University Press).

National Coal Board (no date), *High Technology in Coal* (London: NCB).

National Economic Development Office (1975) *Why Group Technology?,* (London: HMSO).

National Economic Development Office: Process Plant Economic Development Committee (1979, 1981) *Technology Prospects in the Process Industries,* 2 vols (London: NEDO).

National Economic Development Office (1980) *Focus on Engineering Craftsmen: Studies of retention and Utilisation* (London: NEDO).

Nelson, D. (1975) *Managers and Workers: Origins of the New Factory System in the United States, 1880–1920* (Madison: University of Wisconsin Press).

Nelson, D. (1980) *Frederick Taylor and Scientific Management* (Madison: University of Wisconsin Press).

Nelson, R. (1959) 'The Simple Economics of Basic Scientific Research', *Journal of Political Economy, 67,* pp. 297–306.

Nichols, T. and Beynon, H. (1977) *Living with Capitalism: Class Relations and the Modern Factory* (London: Routledge & Kegan Paul).

Noble, D. F. (1979) 'Social Choice in Machine Design: the Case of Automatically Controlled Machine Tools' in Zimbalist (1979).

North West Industrial Development Association (1983) *The Medical Equipment Industry: North West England* (Manchester: NORWIDA).

Northcott, J. (1980) *Microprocessors in Manufactured Products* (London: Policy Studies Institute).

Northcott, J. and Rogers, P. (1984) *Micro-electronics in British Industry: the Pattern of Change* (London: Policy Studies Institute).

Nygaard, K. and Fjalestad, J. (1981) 'Group Interests and Participation in Information System Development', in Organisation for Economic Co-operation and Development (1981) *Information Computer Communications Policy 5: Micro-electronics Productivity and Employment* (Paris: OECD).

O'Brien, R. (1969), *Machines* (New York: Time–Life Books).

Organisation for Economic Co-operation and Development (1979) *Structural Determinants of Employment and Unemployment* (Paris: OECD).

Organisation for Economic Co-operation and Development (1984) *Industrial Robots: Their Role in Manufacturing Industry* (Paris: OECD).

Palloix, C. (1976) 'The Labour Process: from Fordism to Neo-Fordism' in Conference of Socialist Economists (1976).

Penn, R. (1982) 'Skilled Manual Workers in the Labour Process 1856–1964', in Wood (1982).

Perez, C. (1983) 'Structural Change and the Assimilation of New Technologies in the Economic and Social Systems', mimeo.

Perrow, C. B. (1979) *Complex Organisations: a critical essay,* 2nd edn (Glenview, Illinois: Scott, Foreman & Co.).

Pierce, J. L. and Dunham, R. B. (1976) 'Task Design: A Literature Review', *Academy of Management Review,* vol. 1, no. 6, pp. 83–97.

Pollard, S. (1969) *The Development of the British Economy 1914–1967*, (London: Edward Arnold).

Pollard, S. (1981) *Peaceful Conquest: the Industrialisation of Europe 1760–1970* (Oxford: Oxford University Press).

Pollert, A. (1981) *Girls, Wives, Factory Lives* (London: Macmillan).

Porat, M. U. (1977) *The Information Economy*, 9 volumes (Washington, DC: Department of Commerce).

Production Engineer (1981) 'ASP is Alive and Well but Going in a Different Direction', *Production Engineer*, vol. 60, no. 4, pp. 17–21.

Radice, H. (1984) 'The National Economy: a Keynesian Myth?', *Capital and Class*, no. 22, pp. 111–40.

Ranson, G. M. (1972) *Group Technology* (New York: McGraw-Hill).

Rezler, J. (1969) *Automation and Industrial Labour* (New York: Random House).

Riddell, P. (1983) *The Thatcher Government* (London: Martin Robertson).

Rinehart, J. W. (1978) 'Job Enrichment and the Labour Process', mimeo.

Robertson, J. A. S. *et al.* (1982) *Structure and Employment Prospects of the Service Industries* (London: Department of Employment).

Roeber, J. (1975) *Social Change at Work* (London: Duckworth).

Rogers, W. *et al.* (1979) 'Computer-aided Medical Diagnosis', *International Journal of Biomedical Computing*, vol. 10, pp. 267–89.

Rohatyn, F. G. (1984) *The Twenty-Year Century: Essays on Economics and Public Finance* (New York: Random House).

Rose, M. (1975) *Industrial Behaviour: Theoretical Developments since Taylor* (London: Allen Lane).

Rosenberg, N. (1976) *Perspectives on Technology* (Cambridge: Cambridge University Press).

Rosenberg, N. and Frishstack, C. (1983) 'Long Waves and Economic Growth: a Critical Appraisal', paper presented to workshop on 'Innovation, Design and Long Cycles in Economic Development', Royal College of Art, London.

Rosenbrock. H. (1979) 'The Redirection of Technology', paper given at International Federation of Automatic Control symposium, Bari, Italy.

Rothschild, E. (1973) *Paradise Lost: The Decline of the Auto-industrial Age* (London: Allen Lane).

Rothwell, R. and Zegveld, W. (1979) *Technical Change and Employment*, (London: Frances Pinter).

Rowthorn, R. (1976) 'Mandel's Late Capitalism' *New Left Review*, no. 98, pp. 59–83.

Sabel, C. (1982) *Work and Politics* (Cambridge: Cambridge University Press).

Schmookler, J. (1966) *Invention and Economic Growth* (Cambridge, Massachusetts: Harvard University Press).

Shaiken, H. (1980) 'Computer Technology and the Relations of Power in the Workplace', discussion paper (Berlin: International Institute for Comparative Social Research).

Shepard, J. (1971) *Automation and Alienation* (Cambridge, Massachusetts: Massachusetts Institute of Technology Press).

230   *Bibliography*

Shutt, J. and Leach, B. (1984) 'Technical Change in the Food Industry: the Impact of the Ishida Computer Weigher' in Marstrand (1984).

Simons, G. L. (1980) *Robots in Industry* (Manchester: National Computing Centre).

Sims, R. (1982) '1982 – The Year of the FMS', *Production Engineer*, February, pp. 33–6.

Sleigh, J. *et al.* (1979) *The Manpower Implications of Micro-electronic Technology* (London: HMSO).

Smith, M. R. (1977) *Harpers Ferry Armory and the New Technology*, (Ithaca, New York: Cornell University Press).

Soete, L. and Dosi, G. (1983) *Technology and Employment in the Electronics Industry* (London: Frances Pinter).

Softley, E. (1984) 'Women and Technical Change in the Office: Word-processing, Typing and Secretarial Labour', unpublished Ph.D. thesis, University of Manchester.

Sohn-Rethel, A. (1978) *Intellectual and Manual Labour* (London: Macmillan).

Solow, R. (1957) 'Technical Change and the Aggregate Production Function', *Review of Economic Statistics*, vol. 39, pp. 312–20.

Sorge, A., Hartmann, G., Nicholas, I. and Warner, M. (1981) 'Micro-electronics in the Workplace: Unity and Diversity of Work under CNC in Great Britain and West Germany' (Berlin: International Institute of Management).

SPRU Women and Technology Studies (1982) *Micro-electronics and Women's Employment in Britain* (Sussex: Science Policy Research Unit).

Srivastva, S. *et al.* (1975) *Job Satisfaction and Productivity* (Kent, USA: Kent State University Press).

Steffens, J. (1983) *The Electronic Office: Progress and Problems* (London: Policy Studies Institute).

Stokes, W. P. C. (1982) 'Computers in Mining: Progress and Pitfalls', *The Mining Magazine*, December, pp. 560–3.

Stonier, T. (1983) *The Wealth of Information: A Profile of the Post-industrial Economy* (London: Thames Methuen).

Strassman, P. A. (1982) 'Information Technology and Organisations', paper presented to Information Technology '82 Conference, London, December.

Street, D. (1982) lecture given to Manchester New Technology Group by computer consultant for Sperry Univac, 24 February.

Stymne, B. (1966), 'EDP and Organisational Structure', *Swedish Journal of Economics*, vol. 68, no. 2, pp. 89–116.

Swords-Isherwood, N. and Senker, P. (1978) *Technological and Organisational Change in Machine Shops* (Sussex: Science Policy Research Unit).

Swords-Isherwood, N. and Senker, P. (1980); *Micro-electronics and the Engineering Industry* (London: Frances Pinter).

Thompson, J. D. (1967) *Organisations in Action* (New York: McGraw-Hill).

Thompson, P. (1983) *The Nature of Work: An Introduction to Debates on the Labour Process* (London: Macmillan).

Thring, M. W. (1980) *The Engineers' Conscience* (London: Northgate).

Thunhurst, C. (1982) *It Makes You Sick: the Politics of the NHS* (London: Pluto Press).

Trades Union Congress (1979) *Employment and Technology* (London: TUC).

United States Congress (1955) *Automation and Technological Change,* hearings before the Sub-committee on Economic Stabilisation, 84th Congress, Washington, DC.

United States National Commission on Technology, Automation and Economic Progress (1966) *Technology and the American Economy,* 6 volumes, Washington, DC.

Utterback, J. M. (1979) in Hill and Utterback (1979).

Walton, R. (1972) 'How to Counter Alienation in the Plant', *Harvard Business Review,* November–December, pp. 70–81.

Walton, R. E. (1977) 'Work Innovations at Topeka: After Six Years', *Journal of Applied Behavioural Science,* vol. 1, no. 3, pp. 422–33.

Werneke, D. (1983) *Micro-electronics and Office Jobs* (Geneva: International Labour Office).

West, J. (1982) 'New Technology and Women's Office Work' in West, J. (ed.) (1979) *Work, Women and the Labour Market* (London: Routledge & Kegan Paul).

Whisler, T. (1970) *The Impact of Computers on Organisations* (New York: Praeger).

Wild, R. (1975) *Work Organisation* (London: John Wiley).

Wild, R. and Birchall, D. W. (1975) 'Job Structuring and Work Organisation', *Journal of Occupational Psychology,* vol. 48, pp. 169–77.

Wilkinson, B. (1981) 'Technical Change and Work Organisation', Ph.D. thesis, University of Aston in Birmingham; published as Wilkinson, B. (1983) *The Shopfloor Politics of New Technology* (London: Heinemann).

Williamson, D. T. N. (1972) 'The Anachronistic Factory', *Proceedings of the Royal Society of London, Series A,* vol. 331, pp. 139–60.

Williamson, O. (1975) *Markets and Hierarchies* (London: Collier-Macmillan).

Withington, F. G. (1969) *The Real Computer: its Influences, Uses and Effects* (Reading, Massachusetts: Addison-Wesley).

Wood, S. (ed.) (1982) *The Degradation of Work: Skill, Deskilling and the Labour Process* (London: Hutchinson).

Wood, S. and Kelly, J. E. (1982) 'Taylorism, Responsible Autonomy and Management Strategy' in Wood (1982).

Woodward, J. (1965) *Industrial Organisation: Theory and Practice* (Oxford: Oxford University Press).

Yoxen, E. (1983) *The Gene Business* (London: Pan).

Zimbalist, A. (ed.) (1979) *Case Studies in the Labor Process* (New York: Monthly Review Press).

Zimmerman, B. K. (1984) *Biofuture: Confronting the Genetic Era* (New York and London: Plenum).

Zwaan, A. van der (1975) 'The Socio-technical Systems Approach: A Critical Evaluation', *International Journal of Production Research,* vol. 13, pp. 149–63.

Index

232